TECHGENESIS

From Flintstones to AI

by
Parker J. Maddox

TECHGENESIS

From Flintstones to AI

CONTENTS

INTRODUCTION

In the grand tapestry of human history, technology stands out as the thread that weaves our past with the present and sets the stage for our future. From the earliest stone tools to the cutting-edge innovations of artificial intelligence, each leap forward has reshaped our societies, our economies, and our day-to-day lives. This book is an odyssey through the myriad inventions that have marked significant turning points in human evolution.

Imagine a world where the concept of "technology" doesn't exist. It's a world where our ancestors roamed the savannas, relying solely on their physical prowess and raw instincts for survival. The dawn of technology began with simple, seemingly mundane innovations— sharpening a stone into a tool, harnessing the power of fire, and later, inventing the wheel. These rudimentary developments were, in fact, monumental shifts. They were the beginnings of our journey from mere survival to purposeful advancement.

Technology's role in history isn't just about the utility of individual inventions, but about the cumulative impact of these advancements on human societies. It is the story of how the mastery of fire allowed our ancestors to cook food, making nutrients more accessible and fostering brain development. It is about how the invention of the wheel led to better transportation and enabled the exchange of goods and ideas across vast distances.

Throughout our journey, certain periods have acted as catalysts, where innovations sprouted not from a single mind but from a collective yearning for progress. These epochs, marked by rapid technological advancements, were often accompanied by corresponding shifts in our social and economic structures. One can look at the Agricultural Revolution, where new farming techniques led to surplus food production, enabling population growth and the rise of civilizations.

As societies grew more complex, so did our tools and technologies. The ancient civilizations of Mesopotamia and Egypt set the stage for writing and monumental engineering feats, respectively, laying down the blueprints of organized societies. They showcased the dual role of technology in both constructing the physical world and shaping the intellectual and cultural landscapes of humanity.

The classical age of Greek and Roman innovations brought another layer of sophistication. Philosophers pondered the nature of the universe, laying the groundwork for scientific inquiry, while engineers constructed enduring structures like the aqueducts and the Parthenon. These achievements are testaments to the synergy of human thought and practical utility, an interdependence that has continued to fuel our progress.

Then came the medieval period, often mislabeled as the "Dark Ages." Far from being a stagnant era, it was characterized by significant advancements in areas like agriculture, which led to population growth, and the creation of mechanical clocks, which revolutionized how we perceive and measure time.

Every invention, every breakthrough, acted like a stepping stone, enabling the next discovery. The Renaissance rekindled interest in ancient knowledge, spurring advances in art, science, and technology, while the Age of Exploration expanded geographical horizons and forged global trade networks. The printing press democratized knowledge, making information more accessible and spreading new ideas like wildfire.

The Industrial Revolution marked another quantum leap, introducing steam engines and factory systems that forever transformed industry, economy, and society. This era was not just about the emergence of machines but also about urbanization and the profound social changes that followed. The shift from agrarian societies to industrialized cities introduced new challenges and opportunities, compelling us to rethink societal structures and labor dynamics.

The interplay between human needs and technological advancements has continued to evolve. Communication breakthroughs, from the telegraph and telephone to modern digital systems, have shrunk our world, making long-distance communication instant and global collaboration feasible.

The arrival of electricity at the turn of the 20th century electrified our daily lives, unlocking a new realm of possibilities. Transportation, too, underwent radical transformation with the advent of automobiles, airplanes, and advanced railways and ships, further knitting together distant parts of the world.

Modern medicine saw revolutionary changes with the advent of vaccinations and germ theory, drastically increasing life expectancy and quality of life. And then came the digital age, revolutionizing information processing with the birth of computers and the subsequent rise of the internet, fundamentally altering how we work, learn, and interact.

The progression into mobile technology, social media, and renewable energy reflects our continuous quest for innovation. Biotechnology and space exploration further expand our horizons, promising to redefine what we consider possible. Robotics and AI are not only streamlining tasks but also raising existential questions about the future of human labor and artificial cognition.

As we venture into the domains of virtual and augmented reality, ethics and societal impacts become ever more pertinent. Each chapter of this book examines these monumental developments, providing a comprehensive understanding of how technology has been and continues to be a cornerstone of human progress. No invention exists in isolation; every technological advance is both a product of its time and a catalyst for future innovations.

This book invites you to reflect on the intricate dance between necessity and ingenuity that has carried humanity from the caves to the cosmos. It's an exploration into how our tools not only solve immediate problems but also open Pandora's boxes of new questions and possibilities. Through understanding our past, we gain insights into the present and catch glimpses of the future, illustrating the timeless narrative of human creativity and resilience.

As we look forward, contemplating the ethical and societal implications of technologies like artificial intelligence and biotechnology, we recognize that each breakthrough carries with it a blend of promise and peril. The challenge and opportunity lie in how we wield these tools. This book underscores the idea that technology, in all its forms, is a reflection of our collective aspirations and dilemmas.

Ultimately, technology's story is our story—an ever-unfolding saga of discovery, ambition, challenge, and triumph. It's the chronicle of how we've continually reshaped our world and, in the process, redefined what it means to be human. This journey of innovation and exploration invites us to consider where we've been, where we are, and, most crucially, where we're headed.

CHAPTER 1:
DAWN OF INNOVATION

As the sun first broke the horizon of human consciousness, our ancestors embarked on a journey that would see the dawn of innovation ignite. This chapter delves into the momentous era when the earliest forms of human creativity and problem-solving came to life. From the rudimentary but effective stone tools that marked the beginning of technological ingenuity to the revolutionary control and use of fire, these groundbreaking achievements laid the foundation for civilization. Such innovations were more than mere survival mechanisms; they catalyzed a wave of intellectual and cultural growth essential for the complexities that characterize our modern world. The dawn of innovation was, indeed, the spark that set humanity on a path of perpetual advancement, continuously reshaping our existence and aspirations.

The First Tools

Our story of innovation begins with an unassuming rock. Simple yet transformative, these rocks were the very first tools our early ancestors used to shape their world. Long before there was the wheel, or even the concept of fire mastery, early hominids discovered that they could alter their environment with the lift and drop of a stone. It was these rudimentary tools that set the foundation for the marvels of technology that would follow. The advent of the first tools marks a significant chapter in the tapestry of human ingenuity and adaptability.

Consider the Acheulean hand axe. Crafted over a million years ago, this tool is a testament to our ancestors' ingenuity. Early hominids didn't just pick up any rock; they selected specific types of stones that suited their needs. Obsidian began its role in human history due to its sharpness and the ease with which it could be chipped into finer points. Flint and quartzite also found their place in the creation of a variety of tools. The art of knapping – striking a stone to produce a tool – is a practice requiring skill and precision. This ability showcases an evolutionary leap in cognitive function, signifying forward planning and intricate motor skills.

With the development of tools came new ways to hunt, gather, and survive. Simple cutting tools allowed the early humans to butcher meat, slice through tough plant fibers, and craft shelters. These tools reduced the risks associated with raw food consumption by enabling cooking, ultimately leading to increased energy availability for brain development. The bond between these primitive implements and survival is undeniable; they were extensions of the human hand, making tasks easier, quicker, and more efficient.

Social structures and cooperation also evolved around tool usage. Working together on large hunting expeditions or collaborative tool-making not only ensured survival but strengthened communal ties. Sharing knowledge about tool creation became a form of early education, passing down skills through generations. It wasn't just about survival; it was about thriving as a species by capitalizing on shared knowledge and teamwork.

However, the importance of these early tools extends beyond mere functionality. They represent a profound shift in how our ancestors interacted with their environment. Tools allowed for the manipulation of surroundings in a way never seen before, essentially providing a means for humans to exert control over nature. This shift is the very essence of technological progress – turning the elements of the natural world into extensions of human will and ingenuity.

Imagine the sense of wonder and discovery when an early human realized a sharp stone edge could cut through animal hide or carve into wood. Each discovery lit the way for countless generations to expand upon these simple beginnings. From scraping meat off bones to carving symbols in stone, the humble tools served as the first canvas for human expression and communication.

While we often marvel at the rapid advancements of the digital age, it's essential to reflect on these ancient milestones with equal reverence. The crude tools of prehistory laid the bedrock for every subsequent leap forward. Think about the vast gulf between a chipped stone and a smartphone. Yet, it's the mastery of these first tools that set humanity on the path to that very gulf.

Moreover, these tools signified more than technological prowess; they marked a cognitive evolution. The leap from utilizing basic hand-held stones to crafting specialized implements indicated early forms of problem-solving, foresight, and planning. It wasn't enough to find a good rock. They learned to shape and modify it, tailoring their environment to meet various needs – an early sign of the innovative spirit that defines humanity.

These tools also paved the way for the development of more complex inventions. As techniques improved, the ability to create finer and more specialized tools emerged. For instance, the creation of blades and points led to spear-making, which revolutionized hunting techniques and allowed for a more diverse diet. The ramifications of these advancements were profound, impacting settlement patterns, social structures, and even the migration paths of early humans.

Another remarkable development was the invention of composite tools. These tools, which combined different materials (e.g., stone heads attached to wooden handles), represented an amalgamation of various skills and resources. Composite tools demanded a more comprehensive understanding of materials and their properties, highlighting a significant leap in cognitive complexity. This ingenuity displayed an early mastery over nature, heralding the sophisticated technologies that would follow in human history.

The evidence of these earliest tools comes from archaeological sites scattered across Africa, Asia, and Europe. Notably, the Olduvai Gorge in Tanzania offers a treasure trove of early stone tools, shedding light on the inventiveness of Homo habilis and Homo erectus. These archaeological findings provide crucial windows into our past, illustrating the universality of tool use across diverse human cultures. They underscore the shared human journey marked by creativity and resourcefulness.

As we journey forward in time, we see that these initial tools were just the starting point. They evolved steadily, becoming more specialized and efficient. So, when pondering the leaps in technology today—from quantum computing to biomedical engineering—it's humbling to realize they're extensions of the first tools created millions of years ago. The same spirit of innovation that inspired an early human to shape a stone still drives us to this day.

The impact of these first tools cannot be overstated. They didn't just serve immediate needs; they laid the groundwork for civilization. They enabled early humans to manipulate their environment in ways previously impossible, setting the precedent for future technological advances. In every chisel mark on a stone or sharpened edge lies the essence of human progress, a testament to our unyielding quest to innovate and transcend our limitations.

In closing, the significance of these first tools goes beyond their physical utility. They are a symbol of human ingenuity, a precursor to the boundless array of technologies that continue to shape our world.

Just as these early humans transformed simple stones into life-altering objects, we continue to transform our ideas into technologies that redefine what's possible. The spark of innovation that began with the first tools continues to burn brightly, guiding us toward future horizons we have yet to imagine.

Mastery of Fire

Few forces of nature have shaped human history quite like fire. Its mastery marks one of humanity's first monumental leaps in technological innovation. Mastering fire was no simple feat—it was a breakthrough that altered the trajectory of our evolution, culture, and development.

Imagine early humans huddled in caves, shivering against the cold and aching from relentless hunger. When they discovered how to control fire, everything changed. Suddenly, the night wasn't as terrifying; the warmth of a controlled flame gave them unprecedented comfort, banishing the relentless chill of the dark hours. Fire's flickering light extended their days, offering opportunities for socializing, storytelling, and planning—activities that shaped communal bonds and paved the way for the complexities of future societies.

But warmth and light were just the beginning. Fire provided a means to cook food, dramatically changing our diet and nutrition. Cooking made food easier to chew and digest, unlocking nutrients that would have been inaccessible otherwise. This simple yet profound change had cascading effects on human evolution; some scientists believe that cooking allowed our ancestors' brains to grow larger and more complex compared to those of their primate cousins. The relationship between fire and brain development remains a tantalizing area of research, hinting at how closely our intellectual leaps are tied to our technological ones.

Beyond physiological benefits, cooking transformed social dynamics. Meals became communal events, nurturing deeper social structures and cooperative behaviors. The hearth became a gathering place, a precursor to modern social constructs and shared experiences. The very roots of civilization can be traced back to groups of early humans sharing stories and food around the fire. Fire gave us a means to not just survive, but to thrive.

The utility of fire extended into new territories as humans began harnessing its destructive and transformative properties. Fire facilitated the clearing of land for agriculture, allowing for larger and more permanent settlements. It also played a crucial role in early metallurgy. By understanding how to use fire to manipulate materials—melting metals from ores and forging tools—we unlocked countless possibilities that propelled us into new technological realms. Metal tools and weapons enhanced our ability to shape the world to our needs and defend ourselves, laying the groundwork for the complex societies and technological systems that would follow.

Moreover, fire had a profound influence on human mobility. With the ability to create controlled fires, our ancestors ventured into colder climates that were previously uninhabitable. This newfound mobility led to the spread of Homo sapiens across various continents, illustrating how control over a single element can have ripple effects that span the globe. Fire allowed us to adapt to and thrive in diverse environments, proving crucial in our quest for survival and dominance.

However, the mastery of fire was not without risks. Early humans had to reckon with the potentially catastrophic consequences of fire mishandling. Uncontrolled fires could destroy habitats, threaten lives, and wipe out crucial resources. Yet, these dangers did not diminish humanity's persistent experimentation and utilization of fire; instead, they fueled a deeper understanding and respect for its power.

Over time, our relationship with fire grew increasingly sophisticated. We learned to use materials like flint and pyrite to create sparks and kindling to nurture those sparks into flames. Techniques were honed, from the simple act of rubbing sticks together to generating friction, to later advancements involving metallic tools and chemical mixtures. Each step represented an accumulation of knowledge passed down through generations, underscoring the collaborative and cumulative nature of human innovation.

In essence, fire can be seen as the bedrock upon which many other innovations were built. It was a catalyst that encouraged experimentation and problem-solving. As we harnessed fire's power, humanity began to understand the principles of cause and effect in more nuanced ways. This era of discovery laid the foundation for scientific inquiry and technological advancement, themes that have continued to define human history up to the present day.

In many ways, fire's mastery was humanity's first scientific triumph. It required observation, trial and error, and eventually, a deep understanding of its properties and behavior. This fundamental grasp of fire's principles marked our first significant engineering success and became a blueprint for future technological endeavors. Every subsequent breakthrough—from the wheel to the internet—owes a debt to this primal mastery.

The ceremonial and symbolic significance of fire also cannot be ignored. Fire rituals have been integral to many cultures around the world, signifying rebirth, purification, and a connection to the divine. Many ancient traditions celebrated fire as a sacred element, embracing its dual nature of creation and destruction. Even today, fire holds a place of honor in contexts ranging from religious ceremonies to the Olympic torch, a testament to its enduring influence.

As we look back, it becomes clear how fire's mastery served as an inflection point in human history. It provided the framework for society, culture, and technological development. It challenged us to adapt, innovate, and evolve, shaping the course of our journey from primitive beings to the architects of complex civilizations. The story of fire is, in many ways, the story of humanity itself: one of resilience, creativity, and an unyielding pursuit of progress.

Reflecting on the compelling narrative of mastering fire, one can't help but feel a sense of connection to those ancient pioneers who first summoned flames from mere sparks. Their willingness to confront the unknown and harness its power laid the groundwork for subsequent generations of innovators. Each flicker of fire symbolizes a spark of human curiosity and the unending quest to transform the world through technology.

In the grand scheme of human progress, fire serves as both a literal and figurative beacon. It guides us through the annals of history, illuminating our achievements and lighting the path for future innovations. By understanding our ancestors' mastery of fire, we gain insight into the profound interplay between human ingenuity and the natural world, reaffirming our collective potential to ignite change and explore new frontiers.

You might say that mastering fire was our inaugural step on the road to technological mastery. It spurred us to think, adapt, and innovate—values that continue to drive us today. In a real sense, every spark of innovation we see today can trace its lineage back to that first transformative flicker that brightened the lives of our distant ancestors.

CHAPTER 2:
AGRICULTURAL REVOLUTION

The Agricultural Revolution marked a monumental shift in human history, transforming societies from nomadic hunter-gatherers into settled farming communities. This epoch saw the dawn of deliberate crop cultivation and animal domestication, laying the groundwork for civilization itself. With the reliable surplus of food, populations could grow and diversify their skills beyond mere survival. Villages emerged, fostering new forms of social organization and trade. This era of agricultural innovation also led to advancements in tools and techniques such as plows and irrigation systems, fundamentally altering human interaction with the environment. In essence, the Agricultural Revolution was a catalyst for societal progress, heralding an era where human ingenuity began to shape the world's landscape in unprecedented ways.

From Hunter-Gatherers to Farmers

In the grand tapestry of human history, few shifts are as transformative as the move from hunter-gatherer societies to agricultural communities. This seismic change, which we now call the Agricultural Revolution, didn't happen overnight; it was an evolutionary process shaped over millennia, with profound effects on human lifestyle, societal structures, and technological advancement.

For tens of thousands of years, our ancestors roamed the earth, subsisting on the food they could hunt, fish, or gather. Their lives were dictated by the rhythms of nature — the changing seasons, the migration of game, and the cycles of plant growth. But this nomadic existence began to change when humans discovered the extraordinary potential of cultivating plants.

The transition to farming wasn't sparked by a single moment of revelation. Instead, it was a series of gradual realizations and incremental experiments. Early humans began to notice that seeds scattered in fertile soil would eventually sprout and grow. Over time, they learned to intentionally plant these seeds, guard the crop, and harvest the bounty. This nascent form of agriculture allowed them a more reliable food source than the unpredictable pursuit of wild animals and foraging.

Why did this profound shift happen? There are many theories. Climate change at the end of the last Ice Age could have created more stable and warmer conditions favorable for plant growth, which encouraged experimentation with agriculture. Some suggest a growing human population put pressure on traditional foraging resources, pushing innovative survival strategies — like farming — to the fore.

One of the first areas where agriculture took hold was the Fertile Crescent in the Middle East. This region, with its rich soils and favorable climate, became a cradle of early agricultural innovation. Archaeological evidence shows that as early as 9000 BCE, humans in this area were cultivating wheat, barley, lentils, and other crops. As their knowledge and techniques improved, these early farmers began to domesticate animals like sheep, goats, and cattle, adding another layer of stability and productivity to their agrarian communities.

It's important to recognize that the shift to farming didn't just change what people ate — it fundamentally transformed how they lived. With the ability to produce surplus food, humans could settle in one place. Permanent settlements grew into villages and eventually cities, leading to the development of more complex societal structures. These early societies laid the groundwork for modern civilization.

Agriculture was the cornerstone upon which cities, states, and eventually empires were built. The production of surplus food allowed for specialization of labor. Not everyone needed to be involved in food production anymore. Some could become craftsmen, merchants, administrators, or warriors. This diversification led to technological innovation and cultural advancements, as people had the time and resources to develop new skills and ideas.

The spread of agriculture was not uniform. Different regions adapted the concept to their local environments, leading to a variety of agricultural practices worldwide. For example, rice farming became prevalent in Asia, while maize (corn) was a staple in the Americas. These regional differences influenced the development of unique cultures and technologies.

However, the transition from hunting and gathering to farming was not without challenges. Early farmers faced numerous difficulties, from soil depletion and crop failure to the labor-intensive nature of farming itself. They developed new tools and techniques to address these issues, such as irrigation to control water supply, crop rotation to maintain soil fertility, and selective breeding to enhance crop yields.

Domestication of plants and animals was another cornerstone of the Agricultural Revolution. By selectively breeding the most productive plants and taming wild animals for human use, early farmers could significantly increase agricultural output. This process of domestication led to the development of staple crops like wheat, barley, rice, and maize, as well as livestock such as cattle, pigs, and poultry, which remain essential components of our diet today.

From Hunter-Gatherers to Farmers, the advent of agriculture was not just a technological advancement but a social and cultural revolution. The ability to produce food surplus allowed for population growth and the establishment of permanent communities. These communities became the hubs of early human civilizations, where trade, invention, and culture flourished.

The Agricultural Revolution was the catalyst for many other technological innovations. The need to manage and store surplus food led to the development of pottery and granaries. Writing systems emerged as a way to record agricultural transactions and manage resources. The construction of irrigation systems required engineering skills and facilitated advancements in mathematics and mechanics.

Ultimately, the transition from a nomadic lifestyle to settled farming communities enabled humans to form the complex societies we know today. It was a leap forward in our ability to manipulate our environment for sustained productivity and a more predictable way of life. As we continue to innovate and adapt, reflecting on this pivotal shift provides valuable insights into our capacity for ingenuity and resilience.

The Agricultural Revolution was a monumental step in human history, marking the beginning of a new era. It underscored the transformative power of technology and innovation, setting the stage for the advanced civilizations that would follow. By understanding how early humans transitioned from foraging to farming, we gain a deeper appreciation for the incredible journey of human progress.

Domestication and Early Techniques

The dawn of agriculture marks a defining chapter in the human saga, marked by the domestication of plants and animals. This shift wasn't a singular event but a prolonged series of experiments that spanned millennia. From the vantage point of a contemporary observer, the journey from foraging to farming represents a quantum leap, accomplished through ingenuity and relentless adaptation.

In the beginning, our ancestors were primarily hunter-gatherers, relying on the bounty of wild flora and fauna. Yet, a critical mass of knowledge was accumulating. Observations about plant cycles, animal behavior, and environmental patterns began to coalesce into a body of pragmatic wisdom. The realization that seeds could be harvested and intentionally sown to yield predictable crops was nothing short of revolutionary. This nascent understanding laid the groundwork for the establishment of settled communities.

The Fertile Crescent, often termed the cradle of civilization, was one of the earliest regions to witness the full blossom of agricultural practices. The domestication of wheat and barley here spearheaded a transformation. But it's important to note, other regions such as the Yangtze River Basin in China, with its rice paddies, and the Mesoamerican locales where maize became a staple, were also cradles of early agriculture.

Domesticating plants was no small feat. Consider wheat, one of the first grains to be cultivated. Early humans selected for seeds that didn't shatter easily upon ripening, a trait detrimental in the wild but immensely beneficial for farming. These selections over successive generations resulted in robust, high-yield crops. This method of artificial selection—where humans, not nature, chose the organisms with favorable traits—was replicated globally in different contexts, with different plants.

Similarly, animal domestication was a critical part of early agricultural success. Initially, humans pursued animals for their meat, but over time, they realized these creatures could offer much more. Sheep and goats, for example, provided milk, meat, and hides. Their taming involved understanding their breeding patterns, social behaviors, and dietary needs. Cattle soon followed, offering labor as well as sustenance. Domesticated animals started serving as farming aides, food sources, and beasts of burden, revolutionizing agricultural efficiency.

Another noteworthy technique was the use of slash-and-burn agriculture or swidden farming. This method, where forests were cleared and the land was cultivated for a few years before moving on to a fresh area, was widespread. While it provided immediate benefits by enriching the soil with nutrients from burned vegetation, it was unsustainable over long periods, necessitating eventual innovations in crop rotation and land management techniques.

The invention and refinement of tools were indispensable to early agricultural practices. Simple digging sticks evolved into more sophisticated plows, initially pulled by humans and later by domesticated animals. The advent of irrigation systems in arid regions was another groundbreaking development. The construction of canals, ditches, and rudimentary dams allowed communities to control water supply, leading to more predictable and abundant harvests. In Mesopotamia and ancient Egypt, such irrigation systems were the bedrock of agricultural prosperity.

Furthermore, early agriculturalists developed polycropping and crop rotation practices. Polycropping involved growing multiple crops in the same space, mimicking the diversity of natural ecosystems and reducing vulnerability to pests and diseases. Crop rotation, where different crops were planted in a sequence over several seasons, helped maintain soil fertility and reduced the risk of depleting essential nutrients.

Seed storage was another critical concern. Granaries and other storage solutions were developed to protect precious seeds from pests, moisture, and theft. These storage methods became more sophisticated over time, ensuring the viability of seeds for subsequent planting seasons. Methods like drying, smoking, and fermenting foods extended their shelf life, aiding communities in bridging lean periods.

The domestication of plants and animals also had profound social and cultural implications. With a more reliable food supply, populations grew, and social structures became more complex. Surplus production allowed for the specialization of labor. People could pursue activities beyond mere subsistence, leading to advancements in crafts, trade, and governance. Villages evolved into towns and cities, centers of culture and innovation.

In summary, the advent of domestication and early agricultural techniques was a cumulative process, shaped by trial and error, observation, and adaptation. These innovations laid the foundational stones for civilizations, influencing social structures, economies, and future technological developments. Humanity's journey from hunter- gatherers to agricultural societies was a transformative epoch, driven by our ancestors' relentless quest for stability and growth.

The ripple effects of these early techniques are still visible today. Modern agriculture, with its complex machinery, genetically modified crops, and advanced irrigation systems, stands on the bedrock laid by those early pioneers. Understanding this journey provides not just historical insight but inspiration, reminding us of the profound capacity for innovation that resides within us all.

CHAPTER 3:
THE WHEEL AND BEYOND

Progression is never a solitary journey; it's the culmination of countless steps. The invention of the wheel stands as one of humanity's most transformative milestones, not merely as a tool but as a foundation for future innovation. Emerging around 3500 BCE, the wheel's initial simplicity belied its capacity to revolutionize transportation, trade, and labor. Its circular design opened a world of possibilities, from enhancing mobility to boosting productivity across various sectors. As the concept spread geographically and culturally, societies began to adapt and refine the wheel to their unique needs, setting the stage for exponential advancements. The wheel isn't just a relic of the past; it's a testament to human ingenuity and a bridge to future technological triumphs.

Concept and Early Uses

The concept of the wheel is often romanticized as one of the most iconic symbols of human ingenuity. Yet, the journey of its invention is steeped in unexpected simplicity and practicality. The need to move heavy objects more easily led to the groundbreaking realization that cylindrical objects can reduce friction significantly more than dragging. Imagine an early human witnessing the effortless way a log of wood rolls down a hill—it was an idea hiding in plain sight. This observational delight rapidly evolved into purposeful experimentation.

The earliest known use of the wheel dates back to around 3500 B.C. in Mesopotamia. It wasn't a mere happenstance; it was the culmination of an iterative process, much like modern engineering feats. Initially, the wheel was one component of the potter's wheel—a horizontal spinning disc used to shape clay. This usage marked a significant technological leap that revolutionized pottery, demonstrating that the wheel's applications could extend beyond simple transport needs.

On the surface, the wheel's concept appears almost self-evident today, but in its historical context, it was a monumental leap. The brilliance of the wheel doesn't merely lie in its circle form; it resides in the fact that it opens up endless possibilities through axial mechanics. This symbiosis between wheel and axle forged the path to the creation of carts and chariots, fundamentally changing human mobility and economic interactions. It wasn't long before these early adopters of wheeled transport realized they were onto something transformative.

The wheel's ability to enhance the transportation of goods and people had far-reaching consequences. By facilitating trade over long distances, wheels helped forge new connections and enabled cultural exchanges that would have been unimaginable without them. Craftsmanship began to flourish on an unprecedented scale as people were no longer restricted by geographical limitations.

But let's pause for a moment and ponder: why did it take so long for the wheel to be conceptualized and utilized? The answer lies in the interplay of necessity, environment, and cognitive leaps. In regions where other forms of transport, such as sledges in snowy terrains, or boats in riverine environments, sufficed, the urgency to develop a wheeled transport system was minimal. This indicates that the wheel's invention was as much about timing as it was about insight.

What's equally fascinating is how wheels weren't merely confined to transportation. They evolved in tandem with other innovations, ultimately influencing critical aspects of daily life and warfare. For example, once the wheel was adapted for use in carts, innovations in agriculture, metallurgy, and even warfare soon followed. This sporadic, yet significant adoption demonstrates how a single invention can act as a catalyst for broader technological advancements.

Equipped with wheels, communities could transport bulk goods, such as grains and construction materials, which accelerated infrastructural development. Imagine ancient builders constructing grand temples and cities with relative ease compared to their wheel-less predecessors. These wheeled innovations laid down the groundwork for advanced urban planning and the formation of early trade networks. At this point, the wheel wasn't just a technological marvel; it became a crucial enabler of human progress.

On another front, the military applications of the wheel were profound. The advent of the war chariot arguably revolutionized ancient warfare. It allowed armies to move quickly and strike with unprecedented speed and force. The wheel became an instrumental force multiplier on the battlefield. Through a synthesis of craftsmanship and strategic emphasis, entire civilizations leveraged wheeled war machines to expand their territories and influence. This exemplifies the human penchant for adapting tools for multifaceted uses, often pushing boundaries in both peaceful and combative contexts.

But let's not overlook the subtler, yet transformative early applications of the wheel outside of warfare and heavy lifting. The potter's wheel, which preceded the transportation wheel by several centuries, revolutionized the art of pottery making. No longer constrained to hand-shaping and slow coiling techniques, artisans were able to produce more standardized and intricate pottery designs. This artistic leap not only made pottery more functional but also more accessible as an everyday household item. Creativity soared as potters experimented with shapes and forms, leading to a rich tapestry of cultural expression through ceramics.

Another remarkable early use of the wheel was in water-raising devices like the ancient Persian water wheel, known as a "noria." These devices harnessed the wheel's rotational force to lift water from rivers for irrigation. Situated along riverbanks, these wonder machines of the ancient world mitigated the human labor needed for agricultural irrigation, enabling societies to cultivate larger and more productive fields. This wasn't simply the application of a tool—it was the beginning of an agrarian transformation that would have lasting impacts on food production and societal growth.

The wheel also changed the dynamics of human labor and collaborative effort. Before wheels, tasks requiring the transport of heavy goods often necessitated large groups of people. With wheels, fewer laborers could accomplish more. This reduction of required labor changed social structures and economies. Specialized professions such as cart-making and road-building emerged, setting the stage for the kind of division of labor seen in later industrial societies.

What about the non-material impacts? Wheels influenced conceptions of time and efficiency. Once wheeled vehicles became more common, the idea of timeliness, schedules, and deadlines started taking root. The wheel had subtly, yet profoundly, begun to dictate human lives, much like how today's digital clocks and calendars govern our time.

In a broader sense, the early uses of the wheel facilitated the convergence of human economies and environments. As collective inventiveness channeled through the wheel, territorial borders effectively shrank. People traveled farther with more goods and made connections that were previously beyond reach. These interactions sowed the seeds for future cultural and technological exchanges, creating a dynamic web of innovation and ideas.

In recounting this journey from concept to early use, we glimpse the continuum of human progress. The wheel didn't just revolutionize transport; it was a pivotal point of intersection for numerous realms of daily life. What started as an observational genius to ease human effort cascaded into a tool for cultural, economic, and technological growth. The initially simplistic device became a linchpin in the evolutionary narrative of human invention, representing more than just a circle in motion—it exemplified a human propensity for transcending immediate needs and envisioning broader possibilities.

As we delve further into the story of the wheel's spread and its deeper impacts in subsequent chapters, remember that this wasn't the end but a beginning. The humble wheel, through its myriad early applications, laid a robust foundation upon which countless other innovations would build, reshaping civilizations and dictating the pace of progress for millennia to come.

Spreading the Innovation

The invention of the wheel was a breakthrough moment in human history. However, what truly cemented its place in our technological canon wasn't merely the concept itself but the ways it spread and adapted across different cultures and regions. One can draw a parallel to modern technological diffusion, where an idea emerges in one part of the world and soon finds its way into the daily lives of millions, if not billions. The journey of the wheel, from its earliest iterations to its ubiquity, is a testament to human ingenuity and adaptability.

Early evidence of the wheel has been traced back to around 3500 BCE in Mesopotamia. These initial wheels were primarily used for pottery, evolving gradually into more complex transportation tools. From there, the wheel's usage diversified and found its way into various other cultures, including those in Central Asia, Europe, and the Indian subcontinent. Each society adapted the wheel to meet its distinct needs, demonstrating a universal principle of innovation: context shapes application.

In Central Asia, where the terrain varied from steppes to mountains, the wheel evolved into different forms. Spoked wheels made way for chariots, which were instrumental in both trade and warfare. The adaptability of the wheel meant that it wasn't a one-size-fits-all invention but rather a concept moldable to specific regional requirements. This adaptability has always been a critical factor in spreading and sustaining innovations.

Europe saw a different but equally transformative use of the wheel. The concept of the cart, a wheeled vehicle, revolutionized not just how goods were transported but also warfare tactics. Roman legions, for instance, used wheeled chariots and carts extensively for logistics, making them formidable in their military campaigns. Beyond war, the wheel played a critical role in ancient Roman infrastructure, facilitating the construction of roads and aqueducts that stood the test of time.

Meanwhile, in the Indian subcontinent, the wheel found its place in both artisans' workshops and spiritual practices. Potters used wheels to craft intricate ceramics, while massive stone wheels were carved for religious rituals and temple decorations. The diffusion of canines and variations of the wheel across such culturally rich regions highlights how technological innovations transcend practical utility and become embedded in cultural and spiritual idioms.

But innovation wasn't limited to geographical confines. The Silk Road serves as a historical analogy for the internet in its ability to disseminate information and technology across vast distances. Through this network, the wheel reached China, where it catalyzed advancements in cartography and road-building techniques. By 200 BCE, Chinese engineers had developed wheelbarrows, which revolutionized agricultural practices and construction projects. The innovation didn't just travel; it morphed and took root, evolving with each new culture it touched.

The idea of 'borrowing' technology isn't new. What's fascinating is how these borrowed technologies are augmented and transformed. One culture's adaptation not only suits local needs but often pushes the boundaries of the original invention. The wheel's spread exemplifies this: whether in its simplest form or its most complex, it embodies an evolution shaped by and for humans.

Furthermore, the wheel indicates a shift in human thought—a transition from simple problem-solving to complex, abstract thinking. The ability to conceptualize a circular object that could facilitate movement expanded human cognitive horizons. It marked a step toward grasping concepts of physics and mathematics that would become foundational in later technological innovations.

This cross-cultural dissemination of the wheel set a precedent for future inventions. Much like the Renaissance borrowed heavily from the knowledge kept alive in the Islamic Golden Age, or how modern tech hubs around the world today feed off each other's ideas, the spread of the wheel was an early demonstration of humanity's insatiable quest to learn, improve, and innovate.

In today's world, the wheel's legacy is omnipresent. From the smallest gears in a wristwatch to the massive turbines in modern wind farms, the foundational principles of rotational movement continue to find relevance. However, its significance lies not just in its utility but in its journey. The wheel is more than just a tool; it's a symbol of human connectivity and the shared drive to progress.

Imagine trying to describe the world today without the wheel. From urban transportation to medical devices, its application is virtually endless. This underscores the critical nature of spreading innovations. An idea confined to one corner of the world misses out on its potential to change lives across continents. The spread of the wheel serves as an early but powerful reminder of the importance of knowledge sharing in driving human progress.

Indeed, the wheel wasn't only about transportation. In various parts of the world, it served as a basis for water wheels and grinding mills, fundamentally altering agriculture and food production. It's a reminder that the impact of an innovation can extend far beyond its original intended use. Such secondary applications often arise out of necessity and creativity, embodying the very essence of innovation.

The versatility of the wheel's adoption across diverse regions also brings to light an important facet of technological progress: accessibility. Innovations that are simple yet powerful are the ones that often make the most significant impact. The wheel didn't require rare resources or highly specialized knowledge to replicate; it called for a basic understanding of its function and the ability to seamlessly integrate it into existing frameworks, allowing it to thrive in various contexts.

In looking back, the wheel's expansion teaches us valuable lessons about the future. As new technologies emerge, from artificial intelligence to quantum computing, their real value will be measured by how broadly and effectively they can be disseminated and adapted. The path of the wheel offers both a blueprint and an inspiration for spreading groundbreaking innovations in ways that benefit all of humanity.

CHAPTER 4:
ANCIENT CIVILIZATIONS

As human societies transitioned from nomadic lifestyles to settled communities, the intricate tapestry of ancient civilizations began to take shape. These early societies, from Mesopotamia to Egypt, became cradles of innovation, setting the stages for advancements that would echo through millennia. Mesopotamia, often dubbed the "Cradle of Civilization," bestowed upon the world the gift of writing, transforming communication and record-keeping forever. Meanwhile, on the banks of the Nile, Egyptian engineering prowess manifested in awe-inspiring architectural feats, such as the pyramids—wonders that still fascinate us today. These civilizations laid the bedrock for systematic governance, sophisticated art forms, and early science. Their contributions forged the blueprint for urbanization, legal systems, and technological ingenuity that propelled human progress forward, making their legacies a fundamental cornerstone in the grand narrative of technology's evolution.

Mesopotamia and the Birth of Writing

In the fertile crescent of Mesopotamia, where the Tigris and Euphrates rivers nourish the land, human civilization took one of its most significant strides forward—the invention of writing. This transformative leap did more than just record transactions or tell stories; it revolutionized human communication and enabled the administration of increasingly complex societies.

Before the discovery of written language, Mesopotamian societies relied on oral traditions and rudimentary record-keeping methods. Simple tokens made of clay were used to record quantities of goods, but these methods had limitations. As city-states grew, so did the complexity of managing resources, legal matters, and religious rites. It became clear that a more sophisticated system was needed.

The Sumerians, who settled in southern Mesopotamia around 3500 BCE, were pioneers in inventing the first fully developed written language known as cuneiform. This script began as pictographs but gradually evolved into a series of wedge-shaped marks pressed into clay tablets using a stylus made of reed. These tablets, once dried, provided a durable medium for keeping records, which could be archived and referred to as needed.

Cuneiform writing was initially used for economic transactions, such as tracking grain supplies or sheep herds. But its adaptability quickly became evident. Administrative, religious, and literary texts soon appeared, highlighting the system's versatility. One of the most famous early examples is the "Epic of Gilgamesh," a poem that offers insights into Mesopotamian society, beliefs, and values.

Writing also played a crucial role in consolidating power within Mesopotamian city-states. Legal codes, such as the Code of Hammurabi, were inscribed on large stone pillars known as stelae, placed in public spaces for all to see. This promoted a sense of order and justice, guided by written laws rather than oral edicts, which could be easily manipulated or forgotten.

Moreover, writing helped propel the leap from prehistory to history. Recorded events, treaties, and edicts allowed for a more accurate transmission of knowledge across generations. Historians today owe much of their understanding of ancient Mesopotamian society to these clay tablets, which have survived the ravages of time.

Education became pivotally linked to writing, as scribal schools appeared to train the next generation of scribes. These were well-respected positions within society, reflecting the importance of written communication. The curriculum included not just language and writing but also mathematics, astronomy, and law, underscoring the broader role of scribes in Mesopotamian culture.

The impact of Mesopotamian writing was not confined to its place of origin. As the Akkadians, Babylonians, and Assyrians rose to power, cuneiform spread throughout the Near East, influencing a vast region. Even as cuneiform fell out of use, replaced by alphabets that were easier to learn and use, its legacy continued to shape the written traditions of subsequent civilizations.

This early advent of writing in Mesopotamia can be seen as the bedrock upon which much of human progress rests. The ability to record and reflect upon ideas, to draft laws and guidelines, and to preserve stories and religious beliefs, is fundamental to the development of complex societies. Writing marks the dawn of a new era where knowledge could be accumulated, refined, and passed down through generations, creating a lasting dialogue with the past.

The Mesopotamians' achievement wasn't merely a technological advance; it was a profound cultural and intellectual leap. It redefined human interaction, social structure, and even the human capacity for abstract thought. While we might marvel at the impressive engineering marvels of the Egyptians or the philosophical insights of the Greeks, it's crucial to remember that all these advancements were enabled by the written word—a gift from ancient Mesopotamia.

Egyptian Engineering Marvels

Spanning over three millennia, ancient Egypt stands as one of the most remarkable civilizations in human history. One could argue that the true essence of ancient Egyptian prowess lies in their engineering marvels, which to this day continue to astound architects, engineers, and historians alike. These feats weren't just about physical structures; they were about ingenious problem-solving and an understanding of materials and techniques that were well ahead of their time.

The towering pyramids of Giza are perhaps the most iconic symbols of Egyptian engineering. Constructed during the Fourth Dynasty, the Great Pyramid of Giza remained the tallest man-made structure for over 3,800 years. Built as tombs for pharaohs, these pyramids required a level of precision and resource management that is almost unfathomable today. Each stone block, weighing several tons, was transported from quarries located miles away, displaying an incredible feat of logistical engineering.

Beyond the pyramids, the Egyptians demonstrated a remarkable ability to manipulate their environment through massive architectural projects. The construction of the temple complexes, such as those at Karnak and Luxor, showed an unparalleled sophistication in temple design and urban planning. These structures were not merely places of worship but also served as cultural and economic hubs, embodying the very heartbeat of the civilization.

Hydraulic engineering was another area in which the Egyptians excelled. The annual flooding of the Nile River, which deposited fertile silt onto the farmlands, was both a curse and a blessing. To harness this natural phenomenon, the Egyptians developed a series of intricate irrigation channels and basins, which ensured that water could be directed and stored for use during the dry seasons. This advanced understanding of water management not only enabled them to support large populations but also allowed for the development of surplus agriculture, which was a cornerstone of their economic stability.

In terms of materials, ancient Egyptians demonstrated incredible ingenuity. Their use of limestone, granite, and sandstone wasn't limited to just structural functions. They also utilized these materials for

31

artistic and decorative purposes, blending functionality with aesthetics. The obelisks, towering stone pillars often inscribed with hieroglyphics, are prime examples of this intersection of utility and beauty. These monuments were often erected in pairs at the entrances of temples, standing as both engineering achievements and symbols of divine connection.

The Egyptians were also pioneers in medical engineering. The medical papyri recovered from tombs and temples offer extensive texts on procedures and treatments that were highly advanced for their time. Surgical tools made from bronze and other metals have been found, suggesting a practice of medicine that was surprisingly sophisticated. Even their approach to mummification can be regarded as a form of early biological engineering, as it involved a deep understanding of the human body and preservation techniques.

Another marvel of Egyptian engineering was the construction of the world's first known dam, the Sadd el-Kafara, around 2800 B.C. While its intended purpose was to control floodwaters, it ultimately failed due to a rare heavy rainfall. Nonetheless, the attempt itself was remarkable and demonstrated their early understanding of hydrology and structural engineering. Their ability to plan and execute such a massive project, even with the limited technology available, still serves as a lesson in ambition and innovation.

Transportation engineering was yet another area where the Egyptians made significant strides. The development and use of boats along the Nile facilitated not only trade but also cultural exchange between different regions of the country. Their boats varied in design, from simple reed boats to large ships built of wooden planks, showcasing a nuanced understanding of materials and buoyancy as well as shipbuilding techniques.

Moreover, the precision with which they aligned their structures is nothing short of extraordinary. The Great Pyramid, for instance, is aligned with the cardinal points of the compass with incredible accuracy. The ancient Egyptians used simple tools like plumb bobs and sighting instruments yet achieved results that modern technology has confirmed were accurate to within fractions of a degree. Such achievements speak volumes about their observational skills and understanding of the natural world.

Egyptian engineers also showed a keen sense of sustainability, even if that term wouldn't have been in their vocabulary. Many of their buildings and monuments were constructed to last, using materials that would stand the test of time. The recycling of materials was also common practice, as stones from older structures were often repurposed for new buildings. This resourcefulness ensured that materials were used efficiently and economically.

In the realm of mechanics, the ancient Egyptians demonstrated early forms of technology that would lay the groundwork for future innovations. The use of simple machines, such as the inclined plane and lever, allowed them to move massive stones and other heavy objects with relative ease. These early mechanical principles would later be formalized in the works of Greek philosophers and engineers but were already being practically applied in Egyptian construction long before.

Of course, the influence of Egyptian engineering extends beyond their own borders. The knowledge and techniques developed by the ancient Egyptians were passed down through traders, scholars, and conquered peoples, eventually influencing other civilizations. The Greeks, for instance, were heavily influenced by Egyptian methods, adopting and adapting various techniques into their own architectural and engineering practices. This diffusion of knowledge highlights the enduring legacy of Egyptian engineering marvels.

To sum it up, the engineering marvels of ancient Egypt are not just relics of the past but testaments to human ingenuity, perseverance, and the relentless pursuit of excellence. These structures and innovations remind us that even in the earliest chapters of human civilization, the seeds of progress and technological advancement were being sown, shaping a legacy that continues to inspire and awe us to this very day.

CHAPTER 5:
CLASSICAL AGE INNOVATIONS

The Classical Age flourished with groundbreaking advancements, setting the stage for centuries of intellectual and engineering prowess. Anchored by the philosophical and scientific inquiries of the Greeks, and further propelled by the pragmatic engineering genius of the Romans, this period birthed a remarkable synergy of thought and skill. Philosophers like Aristotle and Plato pondered the mysteries of existence, laying the groundwork for scientific methodologies, while engineers and architects like Archimedes and Vitruvius translated these theories into tangible innovations. Roman aqueducts and Greek analytical frameworks were just a glimpse of the era's prodigious output, where abstract thought seamlessly merged with practical execution. This age didn't merely innovate; it laid the foundational bedrock upon which countless future technologies would build, affirming the timeless adage that to understand our present and envision our future, we must stand on the shoulders of giants. Through their relentless pursuit of knowledge and perfection, the ancients bestowed upon us the invaluable gift of progress.

Philosophical and Scientific Foundations

The Classical Age, encompassing the span of ancient Greece and Rome, is often celebrated for its profound and enduring innovations. While much of our focus tends to drift towards the grand architectural feats of the time or the engineering marvels that still stand today, it is essential to recognize the philosophical and scientific foundations that enabled such advances. These foundations were not merely the bedrock of technological progress in the Classical Age; they were the very essence of it.

The philosophers and scientists of ancient Greece and Rome were the pioneers of critical thinking and empirical inquiry. They laid the groundwork for a methodical and systematic approach to understanding the world, which served as a significant departure from the mythological explanations that preceded them. This shift had profound implications for the development and application of technology. By seeking rational explanations for natural phenomena, they established the principles of observation, hypothesis, and experimentation that are still fundamental to scientific inquiry today.

Consider the contributions of early Greek philosophers such as Thales, Anaximander, and Pythagoras. These thinkers were among the first to propose that natural phenomena could be explained by laws of nature rather than the whims of gods. Thales' hypothesis that water is the fundamental substance of the world, although ultimately incorrect, was revolutionary for its time. It represented a shift from supernatural explanations to naturalistic and rational ones, which would later influence countless other fields of study.

Yet, it wasn't merely abstract thought that these early philosophers contributed. Pythagoras' mathematical innovations, for example, are still taught in modern classrooms. His discovery of mathematical relationships in musical harmony formed the basis of what we now understand as acoustics. The Pythagorean theorem remains a cornerstone in geometry, ingeniously linking mathematics to practical engineering and architectural applications. This is a perfect illustration of how philosophical inquiry could directly influence technology and the practical world.

Moving forward, the Classical Age saw the emergence of more formalized scientific study. Perhaps nowhere is this more evident than in the work of Aristotle, whose extensive writings covered subjects as diverse as biology, ethics, and physics. Aristotle's meticulous observational techniques and insistence on empirical evidence laid the groundwork for the scientific method. His "Organon," a collection of works on logic and scientific methodology, can be understood as an early guidebook for empirical investigation. Aristotle posited that knowledge should be derived from systematic observation and logical reasoning, an approach that persisted through centuries to come.

The impact of Aristotle's work extended far beyond his lifetime. When the Roman Empire adopted and adapted Greek knowledge, they not only inherited technological blueprints but also the philosophical underpinnings that encouraged continued exploration and innovation. Roman engineers and architects, for instance, were pragmatic in their applications of these principles. Vitruvius' "De Architectura" encapsulates how Roman society integrated Greek principles of mathematics and philosophy into practical technologies, illustrating the essential unity of thought and practice in advancing human capabilities.

Another central figure in this narrative is Archimedes, whose contributions spanned both theoretical and applied sciences. Archimedes' work on understanding levers and pulleys not only elucidated fundamental principles of physics but also led to practical innovations such as the Archimedean screw, a device for raising water. His mathematical insights into volumes and areas of shapes informed later developments in calculus and engineering. Archimedes' blend of pure theory with applied mechanics exemplifies how philosophical inquiry was not divorced from but rather integral to technological progress.

The Classical Age also nurtured the interdisciplinary nature of knowledge. This period demonstrated that philosophy and science were not isolated domains but interconnected ones that could enhance each other. The Lyceum, established by Aristotle, and the Academy founded by Plato, were among the earliest institutions where knowledge was seen as a collective and integrative pursuit. These institutions emphasized a broad-based education that included mathematics, natural sciences, ethics, and logic—believing that understanding the principles of one field could illuminate another.

Moreover, medical science flourished under this philosophical umbrella. Hippocrates, often regarded as the "Father of Medicine," approached the study of health and disease with a methodical and empirical mindset. The Hippocratic Corpus, a collection of texts associated with him and his followers, laid the foundation for modern clinical practice. His idea that disease results from natural rather than supernatural causes was a radical departure from earlier views and opened the door to systematic medical research and treatment. This shift towards empirical evidence and observation in medicine underscored the broader philosophical tendencies of the Classical Age.

It's impossible to discuss the philosophical foundations of this era without mentioning Socrates, whose dialectical method became a cornerstone of Western philosophy. Though not a scientist in the traditional sense, Socrates' insistence on questioning assumptions and seeking rational explanations influenced subsequent generations of thinkers. This critical approach encouraged the kind of skepticism necessary for scientific inquiry—the willingness to challenge established beliefs and explore new possibilities.

Finally, the legacy of these foundations in the Classical Age persisted well beyond their immediate historical context. The Renaissance, centuries later, saw a revival of Classical ideas, epitomized by figures like Leonardo da Vinci and Galileo Galilei. The integration of careful observation, mathematical reasoning, and empirical testing that characterized the scientific work of the Classical Age was reemphasized during the Renaissance, leading to transformative breakthroughs in multiple fields.

In sum, the Classical Age's contributions to technological and scientific progress cannot be fully appreciated without understanding the philosophical and scientific foundations underpinning them. This era was marked by a profound shift towards seeking rational, systematic explanations for the world around us. The thinkers of ancient Greece and Rome laid down principles that have guided centuries of scientific endeavor and technological innovation. By valuing empirical evidence and logical reasoning, they created an intellectual environment where questioning and exploring became the pathways to understanding and progress.

Thus, it's these philosophical and scientific underpinnings that truly animated the technological wonders of the Classical Age. Their legacy has echoed through time, informing and inspiring subsequent generations to push the boundaries of what is known and what is possible. The spirit of inquiry and the quest for understanding, fostered during this seminal period, remain at the heart of human progress and innovation even today.

Greek and Roman Engineering

As we delve into the marvels of Greek and Roman engineering, it's essential to understand the immense impact these ancient civilizations had on the course of human history. Their innovations laid the groundwork for many of the technological advancements we benefit from today. The Classical Age was marked by unprecedented progress in architecture, infrastructure, and mechanical technologies, all driven by the ingenuity of Greek and Roman engineers.

One of the cornerstone achievements of Greek engineering was their mastery of geometry and its application to architecture. The Greeks' deep understanding of mathematical principles is best exemplified in their temples and theaters. The Parthenon, a prime example, stands as a testament to their skill in combining form and function. Its proportions are meticulously calculated, employing the golden ratio to create an aesthetically pleasing and structurally sound edifice. This blend of artistry and engineering is truly remarkable.

In addition to their architectural prowess, Greek engineers also excelled in constructing advanced infrastructural systems. The Greeks developed aqueducts and water supply networks that were crucial for urban planning. The Eupalinos Tunnel on the island of Samos, which was designed to bring water from a source through a mountain, serves as a significant example. It demonstrates not just the Greeks' ability to conceptualize complex engineering projects, but also their use of precise surveying techniques.

The Romans, inheriting and expanding upon Greek knowledge, revolutionized engineering to an even greater extent. Roman engineering feats were characterized by their scale and ambition, often driven by the empire's needs for military expansion and urban development. One of the most iconic contributions of Roman engineering is the extensive network of roads that facilitated efficient movement across the vast Roman Empire. These roads were meticulously designed with layers of materials for durability and effective drainage, many of which have withstood the test of time.

In the realm of aqueducts, the Romans took the concept to new heights. They constructed expansive networks of aqueducts to supply their cities with fresh water. The Aqua Appia, built in 312 BCE, was the first of many such structures. These aqueducts were engineering marvels, utilizing the principles of gravity and the arch to maintain a steady flow of water over long distances. The Pont du Gard in France is a prime example, showcasing the elegance and efficiency with which the Romans approached these projects.

Another significant Roman contribution to engineering was the development of concrete. Roman concrete, or opus caementicium, was a revolutionary material that enabled the construction of durable and monumental structures. The Pantheon in Rome, with its massive domed roof, epitomizes the potential of this material. The dome remains the world's largest unreinforced concrete dome, a testament to the versatility and strength of Roman concrete.

The Romans' engineering ingenuity wasn't limited to grand projects; it also extended to everyday technologies that improved the quality of life. The hypocaust system, an early form of central heating used in Roman baths and villas, exemplifies this. By circulating hot air beneath raised floors and through hollow walls, the Romans could enjoy warm indoor environments, showcasing an impressive understanding of thermal dynamics.

Military engineering was another area where the Romans made significant strides. Roman military engineers were adept at constructing fortifications and siege works. The construction of Hadrian's Wall in Britain illustrated the strategic use of engineering to secure the empire's borders. This wall, stretching over 70 miles, required extensive planning and resources, demonstrating the Romans' ability to mobilize and apply engineering expertise on a large scale.

Greek and Roman engineering achievements were not isolated but rather the result of cumulative knowledge and technological refinement. The Greeks' advancement in theoretical science and mathematics laid the blueprint for practical applications, while the Romans' emphasis on pragmatism and innovation pushed those boundaries further. Together, these civilizations established engineering principles and techniques that would influence countless generations to come.

Both Greeks and Romans shared a curiosity and drive to solve problems through engineering. Archimedes of Syracuse, a Greek mathematician, and engineer, contributed significantly to our understanding of levers, pulleys, and other mechanical systems. His inventions and discoveries, such as the Archimedean screw for raising water, exemplified the application of scientific principles to practical solutions.

Similarly, Vitruvius, a Roman architect, and engineer, documented a wealth of engineering knowledge in his treatise "De Architectura." His work provides invaluable insights into Roman construction techniques, materials, and design principles. According to Vitruvius, good architecture must possess three qualities: firmitas (strength), utilitas (utility), and venustas (beauty). These principles are evident in the remains of Roman infrastructure and public buildings that continue to inspire modern engineering.

Undoubtedly, Greek and Roman engineering set the stage for future technological advancements. Their achievements in constructing resilient infrastructures, developing new materials, and creating efficient mechanical systems underpin many modern engineering practices. From aqueducts ensuring urban water supply to roads facilitating trade and communication, the legacies of Greek and Roman engineers endure in the very fabric of contemporary society.

As we reflect on these ancient accomplishments, it's clear that the spirit of innovation and the quest for knowledge have continually driven human progress. The Classical Age, marked by Greek and Roman engineering triumphs, serves as a profound reminder of our ancestors' relentless pursuit of improvement. By understanding and appreciating these historical innovations, we gain a deeper insight into the foundations upon which our current technological landscape is built.

In essence, the contributions of Greek and Roman engineering are far more than historical footnotes; they are the bedrock of our civilization's journey through technology. Recognizing their influence helps us appreciate the continuity of human ingenuity and its transformative power, inspiring us to harness that same spirit of innovation as we venture into the future. As we advance, let us not forget the profound

lessons and timeless achievements of our Greek and Roman predecessors, whose brilliance continues to illuminate our path forward.

CHAPTER 6:
MEDIEVAL TECHNOLOGY

During the medieval era, technology evolved in ways that profoundly shaped the future of human civilization. Agricultural advancements like the heavy plow and the three-field system dramatically increased crop yields, ensuring more stable food supplies and supporting population growth. The birth of mechanical clocks introduced an unprecedented precision in timekeeping, which not only revolutionized daily life but also laid the groundwork for future scientific endeavors. Water and wind mills harnessed natural forces to perform labor-intensive tasks, sparing human and animal muscle for other purposes and marking a critical step toward automation. These innovations provided the bedrock for the Renaissance, setting the stage for a cascade of technological breakthroughs that would propel society into a new age of discovery and progress.

Improvements in Agriculture

The transformative period known as the Medieval era brought with it significant advancements in agricultural technology. These innovations were crucial, as they directly impacted food production, population growth, and the overall societal structure. During this time, medieval Europe saw inventions and techniques that revolutionized farming, making it more efficient and sustainable. At the heart of these advancements was a series of innovations that changed the landscape of agriculture forever.

One of the most notable improvements was the development of the heavy plow. Unlike the light scratch plows that preceded it, the heavy plow was designed to turn over the dense, clay-rich soils of northern Europe. This plow was typically equipped with a coulter, a plowshare, and a moldboard, which worked in unison to cut through soil, lift it, and turn it over. This mechanism allowed for deeper tilling, ensuring better aeration of the soil and more effective root penetration for crops. The heavy plow was instrumental in transforming previously unworkable land into fertile farmland, spurring agricultural productivity and enabling the cultivation of heavier soils that were once considered barren.

In conjunction with the heavy plow, the introduction of the three-field system marked a significant leap in agricultural efficiency. Prior to this, the two-field system was the norm, where half of the land was farmed while the other half lay fallow. The three-field system divided the land into three parts: one for winter crops, one for summer crops, and the third left fallow. This rotation not only improved soil fertility by combating nutrient depletion but also maximized the usage of arable land, subsequently increasing agricultural output. By rotating crops and leaving land fallow in a controlled manner, farmers could sustain soil productivity and support larger communities.

Another crucial innovation of the Medieval period was the widespread adoption of watermills and windmills. These technologies harnessed natural forces to power various agricultural processes, such as grinding grain, pumping water, and sawing wood. Watermills, in particular, were essential in areas with accessible water sources. They drastically reduced the labor required for grinding grain into flour, a staple food product. The proliferation of mills allowed for higher grain processing capacity and contributed to a more stable food supply. Additionally, windmills became essential in regions where water sources were less reliable. Both mills symbolized the growing relationship between technology and agriculture, showcasing the ingenuity of medieval engineers in their quest to harness renewable energy sources.

Equally transformative was the medieval period's improvement in harness design and animal-powered farming equipment. Innovations such as the horse collar and the tandem harness allowed horses to replace oxen for plowing and transport. Horses, being faster and more efficient than oxen, increased the speed of fieldwork and made it easier to manage larger plots of land. The horse collar distributed the load across the animal's shoulders, preventing choking and allowing them to pull heavier loads without injury. This shift from oxen to horses lowered the time and labor required for plowing fields, leading to enhanced agricultural productivity.

Advances in hoarding and storage methods also played a critical role in medieval agricultural improvements. The construction of granaries enabled farmers to store surplus grain safely, protecting it from pests and spoilage. These storage facilities were indispensable for managing the food supply, especially during harsh winters or times of poor harvest. Effective storage solutions meant communities could withstand periods of scarcity, maintain a consistent food supply, and stabilize local economies. Larger granaries and better-preserved grains directly influenced the population's resilience against famines.

Furthermore, the period saw notable advancements in crop cultivation techniques. The introduction of new crop varieties and the selective breeding of plants with desirable traits led to higher yields and better resistance to diseases and pests. Techniques like crop rotation and green manuring became standard practices that enriched the soil with vital nutrients, reduced weeds, and cut down on pest infestations. These methods were vital for sustaining long-term agricultural productivity and maintaining balanced ecosystems. Improved understanding and management of soil health were key factors in supporting growing populations and larger, more complex societies.

Moreover, the dissemination of agricultural knowledge and best practices also played a significant role during this period. Monastic communities, with their structured and disciplined approach to land management, became centers of agricultural innovation and education. Monks meticulously documented their farming techniques, which included the cultivation of medicinal herbs, advanced gardening methods, and beekeeping. These religious institutions acted as repositories of agricultural knowledge, promoting experimentation and the sharing of best practices with neighboring regions. Their contributions helped standardize and disseminate effective farming techniques, which spread across Europe through trade and communication networks.

A lesser-known yet crucial development was the improvement of irrigation techniques. Although simpler compared to ancient irrigation systems, medieval methods significantly boosted crop production. Farmers utilized ditches, canals, and rudimentary water-raising devices like the noria to irrigate fields, ensuring crops received sufficient water even during dry spells. These irrigation techniques allowed for more consistent crop growth and expanded the range of cultivable land. Access to reliable water sources often determined the success of agricultural practices and was a key factor in the sustainability of medieval communities.

The continuous improvement in tool-making also contributed to the progress in medieval agriculture. Blacksmiths and craftsmen honed their skills in producing more durable and effective tools, such as sickles, scythes, and hoes. These tools made tasks like harvesting, weeding, and soil preparation more efficient, reducing the overall labor required per unit of agricultural output. The enhancement of implements reflected a broader trend towards specialized tools that increased the efficiency of specific agricultural tasks, contributing to overall farm productivity.

Additionally, animal husbandry and livestock management saw significant improvements. Selective breeding practices improved the quality and quantity of meat, wool, milk, and other animal products. Shepherds and farmers became adept at managing flocks and herds more effectively, resulting in healthier livestock and higher yields. These advances ensured a more varied diet and bolstered local economies through the trade of surplus animal products. The symbiotic relationship between crop farming and livestock husbandry meant that both sectors often supported and benefited from each other, creating more robust agricultural systems.

Moreover, the Medieval period was marked by an increase in agricultural specialization and regional diversity. Different areas began to focus on specific crops and products best suited to their environmental conditions, laying the groundwork for trade networks that moved goods between regions. This specialization not only led to higher quality and more efficient production but also enabled the exchange of agricultural knowledge and practices. As farmers mastered particular crops suited to their local climate and soil, overall agricultural productivity continued to rise, fostering economic growth and societal stability.

In conclusion, the medieval era's agricultural improvements were pivotal in shaping the future of farming and societal development. The innovations of the time—ranging from the heavy plow and three-field system to the adoption of mills and irrigation techniques—laid the foundation for future advancements. These developments enabled medieval societies to sustain larger populations, stabilize food supplies, and create a more interconnected world. As we reflect on these technological strides, it's clear that the period's agricultural innovations had lasting impacts that resonate through history and continue to influence modern agricultural practices.

The Birth of Mechanical Clocks

When we think about the medieval era, it's easy to imagine a time of knights, castles, and agricultural life. But beneath this seemingly static surface, profound technological advancements were quietly reshaping the world. Among these advancements, the birth of mechanical clocks stands out as a transformative achievement. This invention didn't just change how people kept track of time; it fundamentally altered their relationship with time itself, ushering in new rhythms and disciplines that would lay the groundwork for the modern world.

Before the advent of mechanical clocks, timekeeping was a crude affair. People relied on sundials, water clocks, and hourglasses, all of which had their limitations. Sundials were useless on cloudy days, water clocks required constant attention, and hourglasses had to be manually flipped. These methods were adequate for the agrarian societies of the time, where life revolved around the natural cycles of daylight and the seasons. However, as cities grew and the demands of commerce, monastic life, and society at large became more complex, the need for a more precise and reliable method of timekeeping became apparent.

The earliest mechanical clocks began to appear in European monasteries during the 13th century. One of the primary motivations was to regulate the precise scheduling of monastic prayers, known as the Liturgy of the Hours. Monks needed a dependable way to divide their day into regular intervals for prayer, work, and study. These early timekeeping devices were often large, elaborate mechanisms housed in monastery towers, signaling the time with bells. While rudimentary by today's standards, they marked a significant leap from earlier timekeeping methods.

The design of these early clocks primarily relied on weights, gears, and escapements—a small but crucial component in the clock's ability to maintain regular intervals. The escapement is essentially the heart of the mechanical clock. It controlled the release of energy from a wound spring or weighted system, which in turn moved the clock's hands at a consistent rate. This innovation allowed for a new level of precision, paving the way for more sophisticated timepieces to follow.

The role of these early mechanical clocks extended far beyond the cloisters. As towns and cities expanded, clocks began to appear in public squares, symbolizing both technological prowess and civic pride. These public clocks often became landmarks, their hourly chimes serving to synchronize the daily activities of the community. Time, once a nebulous concept dictated by the sun, moon, and stars, became tangible and shared by all.

Moreover, the impact of mechanical clocks on society was profound and multifaceted. For one, they introduced a new temporal discipline that was essential for the development of various social and economic activities. Markets, religious services, and daily labor all began to adhere to more predictable schedules. The heightened awareness of time also influenced the philosophical and cultural outlook of the era. Time became a resource to be managed, a concept that deeply resonated with the burgeoning capitalist ethos of the later medieval period.

In terms of engineering, the development of mechanical clocks was a marvel of ingenuity. Crafting these timepieces required a detailed understanding of mechanics, metallurgy, and craftsmanship. Each gear and wheel had to be meticulously crafted to ensure smooth operation. This intricate work fostered a culture of precision engineering and problem-solving that would later contribute to other technological advancements.

Technological innovation rarely happens in isolation, and the rise of mechanical clocks was no exception. They spurred advancements in other fields, such as navigation and astronomy. Accurate timekeeping was crucial for determining longitude at sea, a challenge that would be tackled with increasing success in the centuries to come. In astronomy, precise clocks enabled more accurate tracking of celestial bodies, contributing to the scientific revolution.

The proliferation of mechanical clocks also had democratizing effects. While the earliest examples were grand public installations or prized possessions of the elite, technological advancements quickly made them more accessible. By the 14th century, portable clocks and smaller mechanisms began to appear, spreading the benefits of precise timekeeping to a broader swath of society. This democratization of time helped bridge social divides and fostered a greater sense of shared experience and order.

Despite their mechanical brilliance, these early clocks were not without their flaws. They were susceptible to wear and environmental factors, requiring regular maintenance. Yet, their very existence and the need to maintain and improve them spurred further innovation. Tinkers and clockmakers of the medieval period laid the groundwork for future inventors, who would go on to develop even more accurate and reliable timekeeping devices.

The birth of mechanical clocks also intersected with other facets of medieval life, including religion, education, and commerce. In cathedrals and universities, clocks became tools for organizing daily routines, reflecting a growing appreciation for regularity and precision. This new mindset played a crucial role in the scientific inquiries of the Renaissance, where disciplined observation and measurement were paramount.

One cannot overlook the inspirational aspect of these early mechanical clocks. They were manifestations of human ingenuity and ambition, tangible proof that people could impose order on the natural world. This spirit of innovation and mastery over time would become a defining characteristic of human progress, motivating future generations to explore and invent.

In summary, the birth of mechanical clocks during the medieval period was a milestone that went far beyond the mere measurement of hours and minutes. It was a catalyst for social change, a driver of economic development, and a harbinger of the scientific endeavors that would shape the modern world. By making time a shared and quantifiable resource, mechanical clocks fundamentally altered human perception and interaction, making their invention one of the most significant technological milestones of the medieval era.'

CHAPTER 7:
RENAISSANCE RENAISSANCE

The Renaissance wasn't just a rebirth of art and culture; it was a transformative period that awakened Europe's intellectual and creative faculties. It beckoned a resurgence of ancient wisdom, blending classical knowledge with bold, new scientific inquiries. This era saw the revival of texts long forgotten, sparking curiosity and advancements in anatomy, astronomy, and physics. Artistic geniuses like Leonardo da Vinci and Michelangelo didn't merely create; they innovated, merging art with science in ways previously unimaginable. Cities became incubators of invention, where thinkers like Galileo and Copernicus challenged old paradigms, forever altering our understanding of the universe. The Renaissance acted as a bridge, connecting the medieval world with the burgeoning age of exploration and discovery, setting the stage for the relentless march of progress that would follow in the subsequent centuries. This period proved that when human potential is ignited, barriers crumble, and boundaries are pushed, propelling humanity into new realms of possibility.

Rediscovery of Ancient Knowledge

The Renaissance, spanning roughly from the 14th to the 17th century, represented a profound cultural and intellectual revival. A key aspect of this rebirth was the rediscovery of ancient knowledge—texts and ideas from classical antiquity that had been preserved through various means, often by scholars in the Islamic world, and reintroduced to Europe. This reclamation of lost wisdom was nothing short of a catalyst, sparking an era characterized by tremendous innovations in art, science, and philosophy.

Centuries before the Renaissance, ancient texts brimming with philosophical and scientific insights lay forgotten or inaccessible in the arcane libraries of monasteries and distant lands. Greek and Roman thoughts had profoundly influenced early European culture, only to fade amidst the dark ages' tumult and religious upheaval. The rediscovery of works by authors like Aristotle, Plato, Euclid, and Ptolemy reawakened Europe's intellectual curiosity, fostering an appetite for inquiry and exploration that would shape the modern world.

One can't overstate the impact of the Islamic Golden Age on this rediscovery. During Europe's Middle Ages, Islamic scholars preserved and expanded upon ancient Greek and Roman texts. Translators like Al-Kindi and Al-Farabi not only ensured the survival of these works but also provided their own contributions to philosophy, mathematics, and medicine. The reintroduction of these texts into Europe via Spain and Italy during the 12th and 13th centuries, often through Latin translations of Arabic works, was akin to opening a floodgate of knowledge.

Italian humanists such as Petrarch were instrumental in this revival. Petrarch, often called the "father of humanism," scoured neglected monastery libraries for ancient manuscripts. His profound admiration for classical writers instigated a broader movement of manuscript hunting, which led to the assembling and studying of a vast corpus of ancient knowledge. This resurgence fueled the belief that humanity could reach new heights of achievement by reconnecting with its intellectual past.

The role of the printing press, an innovation covered in detail in another chapter, also can't be overlooked. The ability to reproduce texts quickly and accurately allowed for an unprecedented dissemination of ancient knowledge. By the end of the 15th century, printed editions of Greek and Roman classics were widely available, ensuring that these texts could be read and reinterpreted by a burgeoning class of scholars and intellectuals.

In the realm of science, ancient knowledge provided a bedrock upon which Renaissance scholars could build. Consider the work of Copernicus, who revisited and challenged Ptolemy's geocentric model of the universe. Copernicus' heliocentric theory, published just before his death in 1543, was shocking and transformative, eventually leading to a paradigm shift in astronomy and natural philosophy. This was a direct consequence of not only accessing ancient texts but critically engaging with them.

Similarly, in medicine, Renaissance figures like Andreas Vesalius revolutionized the field through their engagement with ancient texts. Vesalius' landmark work, "De humani corporis fabrica," was built upon a foundation laid by Greek physicians like Galen. However, Vesalius didn't merely rely on ancient authority; he verified anatomical knowledge through meticulous dissection and observation, thus merging ancient wisdom with empirical research.

Beyond science, the arts flourished under the influence of rediscovered ancient knowledge. Renaissance artists drew heavily from classical antiquity, emulating and surpassing the techniques of their ancient predecessors. The study of human anatomy and proportion, rooted in ancient writings, allowed artists like Leonardo da Vinci and Michelangelo to achieve levels of realism and expressiveness previously unseen. Their works embodied the Renaissance ideal: a harmonious blend of knowledge and creativity.

Architecture, too, underwent a renaissance sparked by ancient blueprints. Architects like Filippo Brunelleschi studied Roman ruins and Vitruvius' "De Architectura," extracting principles of symmetry and proportion that would redefine the skyline of Renaissance cities.

Parker J. Maddox

Brunelleschi's dome for the Florence Cathedral remains a testament to this synthesis of ancient insights and innovative engineering.

The educational landscape also transformed dramatically. The curriculum of medieval universities, which had centered around theology and scholasticism, began to incorporate the studia humanitatis. This curriculum was based on the study of grammar, rhetoric, history, poetry, and moral philosophy—the so-called liberal arts. These subjects drew heavily on classical texts, fostering a broader and more secular educational experience that honored human achievements and potentials.

Philosophically, the Renaissance sparked a return to classical humanism, celebrating the potential for individual achievement and the glory of human creativity. Thinkers like Erasmus and Sir Thomas More engaged deeply with classical texts, updating ancient ideas to address contemporary issues. The emphasis on human potential and the intrinsic value of human life can trace its intellectual lineage back to Socratic dialogues and Aristotelian ethics.

Political thought, too, was transformed by the rediscovery of ancient wisdom. Niccolò Machiavelli's "The Prince," for instance, drew upon historical examples from antiquity to craft a new, realist approach to statecraft. Machiavelli's willingness to depart from idealized visions of governance in favor of pragmatic strategies was a hallmark of the Renaissance's innovative spirit.

The revival of ancient knowledge was not merely an academic exercise; it had real-world implications, shaping policies, technologies, and ideologies. For instance, the advancements in geometry and mathematical knowledge derived from ancient texts had direct applications in navigation and exploratory endeavors, ultimately feeding into the Age of Exploration.

The fusion of ancient wisdom with new ideas fueled a period of unprecedented innovation that reached every corner of European society. By unlocking the forgotten secrets of their intellectual ancestors, Renaissance thinkers laid down the intellectual groundwork for the modern era, bridging the gap between antiquity and contemporary thought.

Thus, the Renaissance wasn't just a period of rediscovery; it was a reimagining. Ancient knowledge served as a lens through which Renaissance scholars could view their world, and in doing so, they could stand on the shoulders of giants. They didn't just regurgitate the wisdom of the past; they interrogated it, reinterpreted it, and, perhaps most importantly, built upon it to reach new intellectual and artistic heights.

In weaving the ancient with the contemporary, Renaissance thinkers crafted a legacy that endures today. The lessons learned from their fusion of past and present continue to inform the evolving narrative of human progress. By understanding how they rediscovered and repurposed ancient knowledge, we gain insights into our own potential to blend history with innovation, ensuring the continual advancement of human civilization.

This convergence of rediscovered ancient wisdom with Renaissance ingenuity set the stage for many chapters yet to come, driving forward the march of human progress and technological development. It's an enduring testament to the power of knowledge and an inspiring reminder of how looking back can often be the best way to move forward.

Advances in Art and Science

The Renaissance Renaissance, as the name suggests, was a time of rebirth. This era wasn't just a revival; it was an explosion of creativity and discovery across various fields, but none more so than in art and science. This period saw thinkers and creators pushing the boundaries of what was previously thought possible. It marked an intersection where beauty met functionality, and imagination coupled with empirical observation transformed the landscape of human thought and achievement.

Art during this time took on a new dimension. No longer were artists merely crafters of religious icons or regal portraits; they became visionaries and technologists. The mastery of techniques like linear perspective, for instance, allowed artists to achieve stunning levels of realism. Consider the works of Leonardo da Vinci, whose notebooks teem with both artistic concepts and engineering designs. His "Vitruvian Man" isn't just a drawing; it's a study in human proportion, anatomy, and the blending of art and science.

Then there was Michelangelo, whose work on the Sistine Chapel ceiling became a monumental example of artistic prowess, as well as a feat of engineering. The use of scaffolding, the manipulation of form, and the understanding of human anatomy all reflect a deeper scientific awareness. These artists weren't working in isolation; they were influenced by and contributing to the scientific discoveries of their time.

Speaking of science, the period saw radical shifts. It was a time when the foundations of modern scientific inquiry were laid. The emphasis on observational empiricism began to take hold, breaking away from the heavily scholastic traditions of the Middle Ages. Figures like Galileo Galilei and Nicolaus Copernicus were instrumental in this revolution. Copernicus's heliocentric theory, which placed the Sun at the center of the universe instead of the Earth, shattered long-held beliefs and faced strong opposition but eventually changed how we perceive our place in the cosmos.

It's impossible to discuss Renaissance science without mentioning the revolution in anatomical studies. Andreas Vesalius's "De humani corporis fabrica" was groundbreaking. For the first time, human anatomy was depicted with such precision that it became an essential reference for both artists and physicians. The detailed illustrations in this work bridged the gap between art and the emerging science of human anatomy, demonstrating a symbiotic relationship that benefited both fields.

Even the tools and instruments of this period were marvels of both beauty and functionality. The development of improved lenses and the invention of the telescope and microscope opened up new worlds to human sight and understanding. Galileo's telescopic observations provided evidence for the Copernican system, while Hans Lippershey's work on the primitive refracting telescope marked the beginning of an entirely new scientific discipline.

Another area where art and science intertwined was in architecture. Advances in engineering and the rediscovery of Vitruvius's texts allowed architects to design and construct buildings that were not only structurally sound but aesthetically exquisite. Filippo Brunelleschi's dome for the Florence Cathedral is a case in point. The use of mathematical principles to solve architectural problems was revolutionary.

In painting, the study of light and shadow, or chiaroscuro, added depth to two-dimensional surfaces, bringing paintings to life in ways previously unimagined. This technique had its roots in scientific exploration of optics and human vision. Artists like Caravaggio and Rembrandt would later push this technique to its zenith, creating works that were not just visually stunning but also psychologically profound.

Moreover, the Renaissance era witnessed the birth of modern scientific literature. Thinkers like Sir Francis Bacon pioneered the scientific method, championing a systematic approach to experimentation and observation. This new approach encouraged others to question, to hypothesize, and to test—a critical departure from relying solely on classical texts and scholastic teaching.

The period also saw advances in mathematics, which played a crucial role in both art and science. The application of mathematical principles in perspective drawing fundamentally changed how artists conceptualized space and form. Simultaneously, in the realm of pure science, figures like Johannes Kepler used mathematical laws to describe planetary motion, marking a significant leap in our understanding of the natural world.

Alchemy, a precursor to modern chemistry, also flourished during this period. While often dismissed today as pseudoscience, alchemy represented an early attempt to understand the building blocks of matter. Alchemists like Paracelsus began to introduce systematic observation and experimentation into their practices, unknowingly laying down the rudimentary principles of modern chemistry.

The printing press, though primarily a technological innovation, had profound impacts on both art and science. For the first time, knowledge could be disseminated widely and rapidly. This not only democratized learning but also accelerated the diffusion of new ideas. Scientific discoveries and artistic innovations were no longer confined to isolated locales but could inspire and be scrutinized by people continents apart.

Exploration of the human form wasn't limited to visual arts; it extended into literature and philosophy. Humanism, a Renaissance cultural movement that turned away from medieval scholasticism and revived interest in ancient Greek and Roman thought, emphasized the potential of human achievement. Thinkers like Erasmus and Thomas More wrote about the human condition, morality, and governance, drawing from classical sources while injecting contemporary insights.

In music, composers like Johann Sebastian Bach and Ludwig van Beethoven began to experiment with new forms and structures, integrating mathematical precision with emotional depth. Although their greatest contributions would come slightly later, the groundwork laid during the Renaissance set the stage for the golden age of classical music.

This era also beckoned the polymaths—individuals whose knowledge spanned multiple fields, seamlessly integrating art and science. Italian poet Petrarch emphasized scholarly love and knowledge expansion, while Sir Isaac Newton, often hailed by future generations, epitomized the synthesis of artistic intuition and scientific rigor.

The spirit of Renaissance continues to inspire. Its advancements in art and science serve as compelling reminders of what happens when creativity meets critical thinking, when observation fuels imagination, and when questioning the status quo leads to breakthroughs. This era established a legacy of interdisciplinary innovation, showing us that profound progress often lies in the symbiotic blend of diverse fields, urging us to pursue wisdom and beauty in unison.

CHAPTER 8:
AGE OF EXPLORATION

As the Renaissance reignited a passion for knowledge and discovery, humanity embarked on one of the most transformative periods in history: the Age of Exploration. This era was marked by a surge in navigational advancements, allowing explorers to traverse vast oceans and chart new territories with unprecedented accuracy. Innovations like the astrolabe and the magnetic compass revolutionized sea travel, enabling daring mariners like Christopher Columbus and Vasco da Gama to connect the world in ways previously unimaginable. The establishment of global trade networks fostered cultural exchanges and the flow of goods, ideas, and technology across continents, laying the groundwork for our modern interconnected world. It was an age characterized not just by the thrill of discovery but by the forging of new paths that would forever alter the course of human civilization, a testament to our unyielding desire to push beyond the horizon in pursuit of progress and understanding.

Navigational Tools

As Europe stood on the brink of the Age of Exploration, one of the crucial challenges faced by explorers was the daunting task of navigating the vast and unpredictable seas. The success of these ambitious voyages largely depended on the effectiveness and reliability of various navigational tools. These instruments were not just mere physical objects; they were technological marvels that captured an extensive history of human ingenuity and advancement.

Chief among these tools was the magnetic compass, an invention that revolutionized maritime navigation. Originating from China, the compass employed a magnetized needle that aligned itself with the Earth's magnetic field, always pointing towards magnetic north. This simple yet effective device allowed sailors to orient themselves even when the skies were overcast or the stars were not visible. By the 13th century, the compass had found its way to Europe and started to become indispensable for navigation in the open seas.

The astrolabe was another essential navigational instrument used during the Age of Exploration. It allowed sailors to determine their latitude by measuring the angle between the horizon and a celestial body, traditionally the sun or a specific star. The astrolabe, with its intricate brass and finely engraved scales, embodied the confluence of art and science. It became a cornerstone for navigators, ensuring that voyages to far-off lands were not just hopeful endeavors but carefully calculated expeditions.

Then, there was the cross-staff, which sailors used to measure the altitude of the sun or stars above the horizon. With its horizontal bar sliding along a calibrated vertical staff, the cross-staff provided a rudimentary but effective means of gauging latitude. Despite its simplicity, it required skill and practice to use accurately, as the measurements were typically taken from the pitching decks of ships swaying in the open ocean.

Later on, the mariner's astrolabe made an appearance—an adaptation of the standard astrolabe but designed specifically for use at sea. It was more robust and less affected by the ship's movement, allowing for more reliable readings. This instrument, forged from sturdy metals, was heavier and more angular than its terrestrial counterparts, built to withstand the rigors of a marine environment.

As the demand for more precise navigation grew, the quadrant, another astronomical instrument, came into use. It consisted of a quarter-circle panel that helped sailors measure the altitude of celestial bodies more accurately. The quadrant's edge was meticulously marked with degrees, providing a reliable way to determine latitude. Its design was continually refined, leading to more sophisticated versions that increased the reliability of navigational readings.

However, accurate navigation did not rely solely on the stars and magnetic fields. The development of better maps and charts was equally vital. Cartographers and explorers exchanged knowledge, the details of coastlines, currents, and prevailing winds meticulously recorded. Portolan charts, with their intricate coastal outlines and detailed navigational instructions, became invaluable resources. Compiled from the collective experiences of generations of seafarers, these charts guided explorers through unknown waters, offering a semblance of certainty in the face of the vast, unseen dangers of the deep.

Another significant breakthrough was the invention of the backstaff, designed to measure the altitude of the sun without having the observer look directly at it. Invented by the English navigator John Davis in the late 16th century, the backstaff allowed for safer and more convenient navigation. By using the shadow cast by the sun rather than its direct light, it reduced the risk of eye damage and made sightings more comfortable and reliable.

With these technological advancements, the dream of reaching distant lands was no longer a matter of chance. Navigational tools empowered expeditions, enabling explorers to traverse vast oceans, map new territories, and establish global trade networks. The sextant, which appeared later, epitomized the pinnacle of navigational technology. By measuring the angle between two visible objects, such as the horizon and a star, it allowed for highly accurate determinations of latitude and longitude. The precision it brought to navigation meant that sailors could pinpoint their location with unprecedented accuracy, significantly reducing the risks of voyages.

The impact of these tools extended beyond individual explorers and voyages. They were instrumental in fostering the growth of global trade networks, linking continents and cultures more closely than ever before. European powers, armed with superior navigational capabilities, could establish colonies and trading posts around the world, shaping the geopolitical landscape for centuries to come. Not only did these instruments facilitate the exchange of goods, but they also enabled the transfer of knowledge, ideas, and innovations across previously insurmountable distances.

The developments in navigational technology during the Age of Exploration serve as a testament to human curiosity and the relentless pursuit of knowledge. These tools were more than just practical devices; they symbolized the quest for discovery and the desire to push the boundaries of the known world. Their evolution reflects a broader narrative of technological progress, showcasing how incremental advancements can lead to monumental shifts in human history.

Reflecting on this era, one can see how the interplay of science, technology, and human ambition created a new paradigm. Navigational tools, in their various forms and functions, were instruments of change. They enabled the Age of Exploration, transforming the way people understood their world and their place within it. The legacy of these tools is enduring, a reminder of how technological innovation can propel humanity towards new frontiers of knowledge and possibility, each tool a small step in the grand journey of human progress.

As we consider the future trajectory of navigational technology, it's worth noting how these early tools laid the groundwork for modern advancements. From GPS systems to satellite-based navigation, the technological lineage can be traced back to those pivotal moments during the Age of Exploration when humanity first truly mastered the art of finding its way across the vast expanses of the globe. The journey from the magnetic compass to modern-day navigation underscores the continuous human endeavor to explore, understand, and connect the world.

In essence, navigational tools were not merely the instruments of exploration but were catalysts for the cultural, economic, and political exchanges that followed. They epitomize the transformative power of technology, highlighting how seemingly simple innovations can have far-reaching impacts on human society. As we delve into the stories of these tools, we uncover deeper insights into the spirit of exploration and the relentless drive that has always propelled humanity forward.

Global Trade Networks

Amid the Age of Exploration, global trade networks emerged as a pivotal force, reshaping economies, cultures, and societies across continents. The web of trade connections that spanned the globe wasn't just about the exchange of goods; it became a catalyst for the transfer of ideas, technologies, and even diseases. Inventors, explorers, and merchants played instrumental roles in establishing routes that connected disparate worlds, knitting humanity in a tighter weave of shared progress and challenges.

The burgeoning maritime advancements of the time were central to this remarkable transformation. Improved ship designs, such as the caravel and carrack, allowed for longer and more reliable sea voyages. Navigational innovations, including the astrolabe and later the sextant, provided sailors with the tools necessary to evaluate their positions with greater accuracy. These advancements made it possible to traverse previously unreachable waters, opening up new avenues for trade and exploration.

European powers, driven by the thirst for wealth and resources, aggressively sought new trading partners and colonies. Portugal and Spain were early trailblazers, with figures like Vasco da Gama and Christopher Columbus—whose voyages were underpinned by groundbreaking navigational technologies—paving the way. Their successful expeditions gave rise to vast colonial empires that reverberated with economic and cultural impacts.

Portugal's establishment of a sea route to India in 1498 marked a seismic shift in global trade dynamics. The route circumvented the traditional Silk Road and the largely Arab-controlled overland routes that had previously dominated Eurasian trade. The Portuguese could now trade directly with India, China, and the Spice Islands (modern-day Indonesia) for spices, silk, and other exotic goods. This significantly reduced costs and increased profit margins.

Spain's discovery of America, prompted by Columbus's voyages, had profound implications. The subsequent colonization and exploitation of the New World led to an unparalleled influx of wealth into Europe, particularly through the extraction of silver and gold. These bounties didn't just enrich the Spanish Empire; they altered the entire economic landscape of Europe and spurred inflation, which some historians argue contributed to the price revolution of the 16th century.

With the rise of these global trade networks, European cities like Antwerp, Lisbon, and later Amsterdam, became bustling hubs of economic activity. The stock exchange in Amsterdam, founded in 1602, represented a novel financial invention, facilitating trade and investment on an unprecedented scale. This burgeoning financial infrastructure allowed for more sophisticated trade and the formation of joint-stock companies like the British East India Company and the Dutch East India Company. These corporations became some of the most powerful economic entities of their time, capable of influencing global trade patterns and even waging war to protect their interests.

In the East, China and India remained central players in global trade networks. China's Ming Dynasty saw the brief but remarkable voyages of Zheng He, who sailed great treasure fleets across the Indian Ocean, extending China's reach as far as Africa's Swahili Coast. The

Chinese silk, porcelain, and tea continued to be in high demand across the globe, fostering a complex network of trade routes that connected Asian suppliers with European consumers. Similarly, India became a pivotal point in the global textile trade, particularly with its high-quality cotton goods that were coveted worldwide.

These global trade networks were not without their dark sides. The transatlantic slave trade, which became integral to colonial economies, brought untold suffering. The forced movement of African slaves across the Atlantic to work in plantations in the Americas was a grievous chapter in human history, driven by the acute demand for labor in new colonial economies. The Middle Passage, the harrowing sea journey endured by enslaved Africans, stands as a stark reminder of the moral costs of unchecked economic ambition.

Another vector of change that came with global trade was the Columbian Exchange, named after Christopher Columbus. This term describes the widespread transfer of plants, animals, culture, human populations, and diseases between the Americas, West Africa, and the Old World in the wake of European exploration and colonization. Staple crops like potatoes, maize, and tomatoes were introduced to Europe, significantly altering diets and agricultural practices. Conversely, Old World crops like wheat and sugarcane found new fertile grounds in the Americas, while diseases such as smallpox had devastating effects on indigenous populations, at times wiping out entire communities.

Overall, these trade networks brought the world's far-flung corners closer together, initiating a more interconnected global economy. This new economic reality required better communication systems, reliable banking, and standardized methods of trade. Letters of credit, bills of exchange, and the use of silver as a global currency started to streamline transactions across continents, laying the groundwork for modern-day financial systems.

The Age of Exploration and the expansion of global trade networks highlight a period where human ingenuity and the thirst for discovery redefined possibilities. Technological advancements in navigation and shipbuilding, driven by an insatiable curiosity and the quest for wealth, were central to this transformation. The resulting global trade networks were complex and multifaceted, bringing advancements and suffering in equal measure. Nonetheless, these networks undeniably set the stage for the modern interconnected world, demonstrating the powerful, often double-edged role that technology can play in our collective human journey.

CHAPTER 9:
THE PRINTING REVOLUTION

The arrival of Gutenberg's press in the mid-15th century was nothing short of a seismic shift in human history. Before this genius invention, the arduous task of hand-copying texts severely limited the spread of knowledge, keeping literacy confined to the privileged few. Gutenberg's movable type brought an unprecedented democratization of information; books became more accessible, varied, and affordable. This revolution accelerated the Renaissance, fueled the Reformation, and laid the groundwork for the Enlightenment, fundamentally altering the intellectual landscape. Suddenly, ideas could traverse continents, igniting innovation and inspiring generations. The Printing Revolution didn't just change the way we consumed information; it was the catalyst that propelled humanity into a new era of shared knowledge and collaborative progress.

Gutenberg's Press

The invention of the printing press by Johannes Gutenberg around 1440 heralded one of humanity's most pivotal moments: the democratization of knowledge. This revolutionary device, uncomplicated in its conception yet profound in its impact, shifted the course of history from the illuminated masculinity of hand-scribed manuscripts to the mass production of written works. Before Gutenberg's press, books were scarce commodities, laboriously copied by hand, making them exorbitantly expensive and accessible only to the elite.

Gutenberg's genius lay in the combination of several existing technologies to create a novel, cohesive system. He adapted the screw press from winemaking and paper production, an idea dating back to the Roman Empire, into a device capable of applying consistent, high pressure to a surface. This was coupled with his creation of movable type—individual letters made from cast metal. Movable type allowed for the reusability of letters and symbols, making the process of setting type for pages far quicker and less labor-intensive than the woodblock printing methods or hand scripting used before.

Furthermore, Gutenberg formulated a uniquely viscous ink, one that could adhere to metal type yet transfer sharply onto paper. This ink was critical in achieving the clarity and durability needed for printed texts. As each component came together, Gutenberg's press gained the potential to produce books with unprecedented speed and uniformity.

It's important to recognize the ripple effects of this innovation. The first major project undertaken with Gutenberg's press was the now-famous 42-line Bible, also known as the Gutenberg Bible. Produced in around 1455, this Latin Vulgate text demonstrated the vast possibilities of the press. With roughly 180 copies created, it was a monumental figure compared to the hand-scribed texts that took scribes years to complete, often resulting in only a single copy.

One fascinating aspect of Gutenberg's Press is how it blurred the lines of craftsmanship and industrial production. On the one hand, the initial setup of each page required careful, artisanal attention. On the other, once set, those pages could be duplicated hundreds or even thousands of times with relative ease. This synthesis of artistry and mass production set a precedent for future technological advances.

The implications of Gutenberg's press transcended the realm of books and scholarship. The printing press stimulated the information economy, disrupting medieval power structures dominated by the clergy and the aristocracy. It fostered an environment where ideas could proliferate without the gatekeeping of religious or governmental authorities. Literature, scientific research, and philosophical musings all began to circulate more freely, igniting intellectual movements.

The Protestant Reformation, for instance, gained momentum partly due to the printing press. Martin Luther's 95 Theses, nailed to the church door in 1517, were rapidly disseminated through the newly established print networks. Similarly, the Scientific Revolution benefited immensely from printed works. When Copernicus published "De Revolutionibus Orbium Coelestium" in 1543, articulating his heliocentric theory, the press ensured that his radical ideas reached a broad audience.

Gutenberg's press also accelerated the spread of literacy. As books became more affordable, increasing numbers of people had access to written materials. Educational systems began to evolve to include a broader segment of society, fostering not just literacy but critical thinking and informed discourse. The rise of a more literate public dovetailed with emergent political structures, laying the groundwork for modern democratic principles by empowering citizens with the knowledge required to participate meaningfully in governance.

As the centuries progressed, the basic principles of Gutenberg's press evolved, yet the core ideas persisted. Innovations like the rotary press and linotype machine built upon his methods, creating even more efficient means of producing texts. These technological successors continued to lower the barriers to information, catalyzing movements like the Enlightenment and giving rise to modern journalism.

In exploring the impact of Gutenberg's press, one can't underestimate its role in shaping our contemporary world. The efficiencies it introduced in reproducing written content underpin today's digital and data-driven society. Though digital text and e-books may seem light-years away from Gutenberg's movable type, they're part of the same continuum of striving to make knowledge more accessible. In essence, the spirit of Gutenberg's innovation lives on, continuing to influence our interactions with information and each other.

While it's intriguing to delve into the technical aspects and historical significance, let's not lose sight of the inspirational nature of Gutenberg's story. Here was an individual driven by the unwavering belief that the dissemination of knowledge could transform society. It's a tale of perseverance against odds, of keen observation, and of harnessing available technologies to create something transcendent.

Gutenberg's journey wasn't without its setbacks. For instance, his partnership with financier Johann Fust didn't end amicably. Fust ultimately sued Gutenberg, leading to a lawsuit that resulted in Gutenberg losing control of his press and his equipment. Yet, despite these challenges, the ripple effects of his invention were unstoppable. The seeds planted by his press continued to flourish, influencing countless facets of human development.

It's worth examining our own time and acknowledging parallels to Gutenberg's era. Today, as we stand amidst the rapid advancements of the digital age and artificial intelligence, there's a bit of Gutenberg in every byte of data we share, in every piece of content we consume. Like him, we navigate an unprecedented flood of information, continuously shaping and being shaped by it. In this voyage through the annals of technological progress, Gutenberg's press serves as a compass, reminding us of the power and responsibility inherent in our quest to democratize knowledge.

Indeed, Gutenberg's press was more than a machine; it was a catalyst for an intellectual renaissance, a beacon for the democratization of knowledge, and a harbinger of an age where information flowed freely, unbounded by the constraints of time and geography. As we marvel at advancements in communication, transportation, and beyond, let's remember the ink-stained hands that set it all in motion, one metal letter at a time.

The Printing Revolution begot by Gutenberg's innovation not only altered the trajectory of human history but also illuminated the path forward—a path where access to information and knowledge signifies empowerment, and empowerment, in turn, fuels progress.

Impact on Literacy and Knowledge

The invention of the printing press by Johannes Gutenberg in the mid-15th century wasn't just a technological breakthrough; it was a seismic event that fundamentally altered the landscape of knowledge and literacy. Before the arrival of the printing press, books were painstakingly hand-copied by scribes, often taking months or even years to produce a single volume. This laborious process made books scarce and prohibitively expensive, accessible only to the elite strata of society.

With Gutenberg's press, the replication of texts became exponentially faster and cheaper. For the first time, books could be produced en masse, paving the way for a democratization of knowledge. The conduit through which information flowed was no longer restricted to monasteries and wealthy scholars. Suddenly, anyone with the means to purchase or borrow a book could access an ever-expanding pool of information. This paradigm shift empowered the average person to seek out knowledge independently, fueling the growth of personal libraries and the rise of public libraries.

The spread of printed materials had an incredible impact on literacy rates across Europe. Prior to the printing revolution, literacy was largely confined to the clergy and a few privileged nobles. The newfound availability of books revolutionized education, enabling broader segments of society to attain literacy. Schools began to incorporate printed texts into their curricula, and the rise of secular universities promoted learning beyond theological studies. This gradual increase in

TechGenesis

literacy rates fueled a burgeoning middle class capable of critical thinking and innovation.

Beyond merely improving literacy, the printing press also revolutionized the way knowledge was disseminated and consumed. Scientific discoveries, philosophical treatises, and literary works could now be shared more widely and accurately. For example, the works of scientists like Nicolaus Copernicus and Isaac Newton reached a much broader audience, accelerating the scientific revolution. Scholars could now correspond and share their work with peers across great distances, fostering an unprecedented level of intellectual collaboration and debate.

The impact on literacy and knowledge cannot be overstated when considering its role in the Reformation. Martin Luther's "Ninety-Five Theses," nailed to the door of the Wittenberg Castle Church in 1517, were quickly printed and circulated widely. This not only accelerated the spread of Reformation ideas but also demonstrated the power of the printing press as a tool for societal and religious change. The ability to disseminate alternative viewpoints swiftly weakened the centralized control of information traditionally held by the Church and other authorities, sparking debates that would shape the modern world.

The printing press also had profound implications for language and the standardization of texts. Before its invention, regional dialects were prevalent, and there was little to no standardization of spelling or grammar. The widespread distribution of printed works began to normalize languages and standardize spelling and grammar. This, in turn, made communication clearer and more efficient, aiding in the development of national identities—particularly crucial during the age of nation-states.

The accessibility of printed material allowed for the preservation and dissemination of ancient texts as well, aiding in the Renaissance movement. Rediscovered works from Greek and Roman authors were printed, making the wisdom of the ancients available to a new generation of thinkers. The accessibility of these texts inspired advancements in various fields, including art, science, and literature, echoing through the centuries as the basis for much of Western thought.

Furthermore, the printing revolution played a pivotal role in shaping public opinion. For the first time, pamphlets, newsletters, and later newspapers created a forum for public discourse. The proliferation of printed material meant that ideas could be debated in the public sphere, forming the bedrock of what would eventually become modern democratic societies. This new platform for dialogue allowed not only scholars and politicians to express their views but also gave a voice to those who had previously been marginalized.

Even in realms of fiction and storytelling, the printing press left its mark. The novel as a literary form gained widespread popularity during this time, expanding the horizons of what could be achieved in storytelling. Writers like Miguel de Cervantes and William Shakespeare reached an audience they might never have known, influencing literature for generations. The rise of fiction also played a role in developing empathy and social awareness among readers, fostering a more interconnected and empathetic society.

The academic world saw immense benefits as well. With access to a plethora of texts, scholars could engage in comparative studies, cross-referencing works from different regions and epochs. This practice cultivated a more nuanced and comprehensive understanding of various subjects. The ability to easily reproduce scholarly works also meant that educational institutions could standardize curricula, further formalizing and improving education systems.

The ripple effects of the printing press reached even into the realm of economics. As literacy rates climbed, so did economic opportunities. Educated individuals could engage in better job prospects, and a more knowledgeable workforce contributed to overall economic growth. The ability to read instructions, grasp contracts, and understand financial documents made businesses more efficient and fostered the growth of commerce.

In summary, the printing revolution was a transformative period that reshaped the world in immeasurable ways. By making books and other printed materials more accessible, it democratized knowledge and fueled an increase in literacy rates. The rapid dissemination of ideas led to significant advancements in science, religion, politics, and culture, laying the groundwork for many aspects of contemporary life. The impact on literacy and knowledge created a virtuous cycle—education led to innovation, which in turn necessitated more education. This cycle continues to drive human progress forward, demonstrating the lasting power of the printing press as one of humanity's most revolutionary inventions.

CHAPTER 10:
INDUSTRIAL REVOLUTION

Arguably one of the most transformative periods in human history, the Industrial Revolution marked the dawn of a new era of relentless progress and profound change. It was a time when steam engines roared to life and factories sprouted up like modern cathedrals to industry's rapid advance. The introduction of mechanized production not only boosted manufacturing capabilities but also sparked a wave of urbanization as people flocked to burgeoning cities in search of work. This seismic shift didn't just reshape the economic landscape; it altered the very fabric of society. Social hierarchies were upended, traditional lifestyles were disrupted, and a new class of workers emerged, laying the groundwork for the modern capitalist economy. The Industrial Revolution stands as a testament to human ingenuity, showcasing how technological advancements can propel us into new realms of possibility and fundamentally alter our world.

Steam Engines and Factories

Stepping into the Industrial Revolution, the development of steam engines and the rise of factories are pivotal moments that reshaped economies, societies, and the very fabric of everyday life. These technological innovations didn't just come out of nowhere; they were built on centuries of accumulated knowledge and effort. Yet, their introduction brought about changes so rapid and profound that the world would never be the same.

The story of the steam engine begins with the need to remove water from coal mines. Early engines, like Thomas Newcomen's atmospheric engine in the early 18th century, were cumbersome and inefficient by later standards. But they served their purpose, pumping water out of mines, allowing deeper seams to be exploited. Newcomen's engine laid the groundwork for what was to come, proving that steam could be harnessed for practical work.

James Watt, often heralded as the father of the steam engine, took Newcomen's invention and brought it to new heights. In the 1760s, Watt's improvements—such as the separate condenser—drastically increased efficiency, making the steam engine a viable power source for a range of applications. It wasn't just for mines anymore; it could drive machinery in mills and factories across various industries. This innovation marked a significant leap in human ingenuity, transforming the way energy was produced and consumed.

Factories powered by steam engines began to sprout across the landscape like mushrooms after a rainstorm. One of the earliest and most notable examples was the cotton mill. Thanks to inventions like the spinning jenny and the power loom, textile production shifted from home-based, manual labor to factory-based, mechanized processes. These factories could produce cloth at unprecedented speeds and in larger quantities. The result was a reduction in the cost of textiles and a boost to the economy, but it also set the stage for complex social dynamics and labor issues that would characterize this new era.

The shift from craft-based production to factory-based manufacturing was more than a technological change; it altered the very essence of work and society. Factories required large numbers of workers, leading to the migration of people from rural areas to burgeoning industrial cities. Thus, urbanization and the growth of factory towns became a hallmark of the Industrial Revolution. Cities like Manchester in England swelled with workers seeking employment, transforming it into an early symbol of industrial power.

Life in these factories, however, was far from idyllic. Workers—including women and children—faced grueling hours, low pay, and unsafe conditions. This period saw the rise of labor movements calling for better working conditions and fair wages. It was a time where the benefits of technological progress were often misaligned with human well-being, leading to social reforms and the eventual establishment of labor laws. The relentless march of progress kept its pace, but so did the struggle for human rights and dignity.

The impact of steam engines extended beyond textile mills. They found applications in transportation as well, most notably in the railways. George Stephenson's Rocket locomotive and the subsequent network of railroads revolutionized the movement of goods and people. Trains facilitated faster, cheaper, and more efficient transport, shrinking distances and opening up new markets. This capability fueled further industrial growth, creating a feedback loop of innovation and expansion.

Another sector profoundly affected by steam power was metallurgy. Factories and foundries equipped with steam hammers and rolling mills could produce large quantities of high-quality iron and steel. This output was essential for constructing the machinery and infrastructure needed to sustain and expand industrial activities. Iron bridges, steel rails, and iron-clad ships became symbols of an industrialized world, showcasing the material power generated by steam engines.

While the technological advancements of steam engines and factories were monumental, they also forced society to grapple with unforeseen consequences. Environmental degradation, air and water pollution, and the stark contrasts of wealth and poverty became common issues. These challenges spurred the growth of new fields of study and policy aimed at mitigating the negative impacts of industrialization. Urban planning, labor rights, and environmental science can trace some of their origins back to this era of explosive growth and its accompanying problems.

The spread of factories wasn't limited to textiles or transportation; virtually every industry felt the ripple effects. Food processing, paper manufacturing, and even chemical production saw transformations that increased efficiency and output. As factories became more specialized, they also became more reliant on a steady supply of raw materials. This dependency further accelerated global trade and colonization, as industrial powers sought to secure these resources from around the world.

Steam engines and industrialization also had profound effects on agriculture, albeit indirectly. Innovations like the steam-powered thresher and tractor later emerged, accelerating farm productivity. While these developments came later, the groundwork was laid during the early Industrial Revolution, illustrating the far-reaching effects of steam power beyond the factory walls.

The ripple effects of steam engines and factories reverberated through culture and society. Literature from the period reflects a spectrum of reactions, from the triumphalism of progress to the stark social critiques found in the works of Charles Dickens and others. These narratives provide a window into the complexities and contradictions of a world rapidly changed by industrialization. The stark realities of factory life contrasted sharply with the promise of a better future made possible by technological progress.

Viewed through the lens of history, the steam engine stands as one of humanity's greatest inventions, embodying both our ability to solve practical problems and the ethical dilemmas that come with such power. The age of steam and factories was a crucible, testing the limits of human ingenuity and resilience. As we look back on this transformative period, it becomes clear that while technology can propel us for-

Parker J. Maddox

ward at unprecedented speeds, it requires a conscious effort to ensure
that progress benefits all of humanity.

Urbanization and Social Changes

The Industrial Revolution stands as a transformative period in human
history, characterized by mechanized production, evolving labor dynamics, and, crucially, urbanization. As factories sprung up in previously rural landscapes, the gravitational pull of job opportunities drew
masses from the countryside to burgeoning cities. The phenomenon of
urbanization during the Industrial Revolution wasn't just about the
physical clustering of people; it marked a seismic shift in social structures and community life.

Before this era, most of the world's population lived in agrarian
landscapes, tethered to the rhythms of the natural seasons. Villages and
small towns were the epicenters of social life, where everyone knew
everyone else. With the advent of industrialization, that close-knit fabric began to unravel. People migrated to cities, where anonymity replaced familiarity, and traditional social ties loosened.

The migration wasn't just about seeking employment. It was also
about escaping the limitations and hardships of rural life. Agricultural
work was seasonal and precarious. Industrial jobs, while often grueling
and monotonous, offered a steadier income. This influx of workers led
to the rapid expansion of urban areas. Cities like Manchester, Glasgow,
and Chicago transformed from small towns into industrial power-
houses almost overnight, embodying the promises and perils of the
new age.

One of the most glaring changes was the living conditions these
new urban inhabitants faced. Cities were ill-prepared to handle the
rapid population growth. Housing shortages and overpopulation led
to cramped, unsanitary living conditions. Tenement buildings, often
poorly constructed, became the norm. These multifamily units were

breeding grounds for diseases, fostering unsanitary conditions that quickly spread illnesses among the dense populations.

Despite these challenges, urban living also brought unprecedented opportunities. The concentration of people facilitated the growth of a new social dynamic. With the rise of factories, a clear division of labor emerged. Jobs became specialized, and a class of industrial workers, skilled artisans, and new managerial roles started to coalesce. This diversification of work laid the groundwork for a complex socio-economic structure.

Urbanization also amplified social stratification. The divide between the wealthy industrialists and the working poor became more pronounced. Factory owners and business magnates amassed fortunes, often living in opulent neighborhoods far removed from the squalid conditions of the working class. This disparity fueled social tensions and eventually led to movements advocating for labor rights and reforms.

The Industrial Revolution's urbanization also drastically reshaped the physical and social landscape of cities. Infrastructure had to evolve to meet the demands of growing populations. Roads were paved, public transportation systems, such as trams and railways, were introduced, and essential services like sewage systems and clean water supplies began to be implemented—albeit gradually and often slowly.

Conversely, the explosion of urbanization brought a plethora of cultural and intellectual boons. Cities became melting pots of ideas and cultures. Public spaces like parks, theaters, and museums started to emerge, offering new forms of social and cultural engagement. Theaters showcased new plays, musicians found audiences in bustling city squares, and intellectual salons became crucibles for debate and discussion on everything from philosophy to science to social justice.

Women, in particular, found new opportunities as they moved into urban centers. Factories provided employment for women, who were previously restricted to domestic duties or agricultural work. Cities also became hubs for the early women's rights movement, as activists gathered to push for suffrage and labor reforms. Although the conditions were far from ideal, urbanization offered a platform for women's voices to grow louder and more influential.

It is essential to understand that this transformation did not occur uniformly across all urban areas. The extent and speed of change varied among different cities and countries, influenced by local economic, political, and social factors. Some cities managed their growth and resources more efficiently, providing better living conditions and public services, while others lagged, exacerbating the difficulties faced by their inhabitants.

The Industrial Revolution's urbanization also had a profound impact on the family structure. Extended families, which had been a cornerstone of rural life, often fragmented as individuals moved to cities for work. Nuclear families became more common, and the traditional patriarchal family model started to erode under the new economic realities. Women and children often worked alongside men in factories, contributing to the family's income but also complicating traditional household roles.

Educational opportunities also expanded, albeit unevenly. As cities grew, so did the need for educated workers to manage and operate increasingly complex industrial processes. This led to the establishment of more schools and eventually public education systems. The spread of literacy and education had long-term implications, fostering innovation and enabling social mobility for some, despite significant barriers that still existed for many, especially the working class and marginalized communities.

Neighborhoods within these industrial cities often became micro-cosms of cultural diversity. Immigrants from various regions and countries settled in urban centers, each bringing their unique traditions, cuisines, and languages. This multiculturalism enriched city life but also introduced new challenges, such as racial and ethnic tensions that sometimes erupted into conflict.

Urbanization during the Industrial Revolution was a double-edged sword. It catalyzed economic growth and provided new opportunities, but it also brought about significant social challenges. It shook the foundations of existing social orders and required societies to adapt rapidly. Through this tumultuous period, human resilience and ingenuity shone brightly, laying down the patterns of urban living that continue to define our world today.

The Industrial Revolution's legacy of urbanization is a reminder of how technological advances can reshape not just industries, but the very fabric of our lives. It underscores the intertwined nature of technological progress and social evolution, offering lessons on the complexities and consequences of rapid change. The impact of urbanization continues to be felt, as modern cities grapple with challenges inherited from this pivotal era, striving to find balance and sustainability in an ever-evolving technological landscape.

CHAPTER 11:
COMMUNICATION BREAKTHROUGHS

The evolution of communication systems marks one of the most significant leaps in human progress, fundamentally transforming how societies interact and share information. The advent of the telegraph obliterated geographical barriers, allowing people to transmit messages over vast distances with unprecedented speed. This innovation laid the groundwork for the telephone, which made real-time vocal communication possible, further shrinking the world and fostering global connectivity. As we transitioned from Morse code to spoken conversations, the stage was set for the development of modern communication systems that encompass everything from radio waves to fiber-optic networks. These technologies have not only revolutionized personal and business communication but have also catalyzed advancements in education, science, and international diplomacy. Each breakthrough in this domain has acted as a stepping stone, leading us to a world where instant, global communication is a reality, reshaping how we perceive and interact with the world around us.

Telegraph and Telephone

The telegraph and the telephone were monumental breakthroughs that revolutionized human communication. They each played pivotal roles in shaping the modern world, transforming how people connected over distances. Telegraphy emerged first, laying the groundwork for the rapid exchange of information, while the telephone built upon this foundation, enabling voice communication over vast expanses. Together, these inventions compressed time and space, fostering an interconnected global society.

Imagine a world where long-distance communication was limited to the speed of a horse or a ship. Messages took days, often weeks, and sometimes even months to reach their destinations. The telegraph changed all that. Developed in the 19th century, it provided an almost instantaneous method to transmit information across long distances using electrical signals and Morse code. Samuel Morse and his collaborators brought the first commercially successful telegraph system into operation in the 1840s. The simple dots and dashes of Morse code sent over wires marked the beginning of a new era.

Even though the telegraph's code was simple, its ramifications were profound. Suddenly, news traversed continents in minutes instead of days. This shift meant quicker response times in trade, politics, and even personal relationships. Wars like the Crimean War and the American Civil War saw the strategic advantage of real-time communication, which contributed significantly to the direction and outcomes of these conflicts. The telegraph reduced the world to manageable dimensions, making long-distance communication not only feasible but also routine.

In the business world, the telegraph was a game-changer. Financial markets were among the earliest adopters, using the technology to share stock prices instantaneously. News agencies sprang into existence, relying on telegraph networks to distribute information rapidly. Through a network of wires, the telegraph enabled the global expansion of businesses, from humble traders to mighty industrial enterprises. It was as though the world shrank, creating an unprecedented level of interconnectedness.

Yet, the telegraph had its limitations. It relied on a rudimentary binary code that required trained operators on both ends. Moreover, messages were often brief and lacked nuance. Enter Alexander Graham

Bell and the invention of the telephone in 1876. While the telegraph used coded signals, the telephone transmitted actual voices, allowing people to communicate as if they were in the same room. Although it took a decade or more for the telephone to become widespread, it fundamentally changed how people interacted.

Alexander Graham Bell's first successful demonstration of the telephone's capability was famously marked by the phrase, "Mr. Watson, come here, I want to see you." These words encapsulated the potential of the telephone: immediate, clear, personal communication, far surpassing the telegraph in richness and immediacy. The telephone didn't just add to the communication landscape; it transformed it. For the first time, people could hear each other's voices across great distances, capturing the tones and inflections that written words and Morse code could not convey.

The telephone's impact extended beyond personal communication. It also revolutionized business operations, enabling real-time discussions and decisions that were previously impossible. Imagine negotiating a trade deal millions of miles apart; the telephone closed that distance and made it feasible. Financial transactions, legal consultations, and medical diagnoses could be conducted over the phone, making operations more efficient and effective.

Despite initial skepticism and technical challenges, the adoption of the telephone grew rapidly. The first telephones were direct lines between two points. Eventually, switchboards and operators became standard, allowing people to connect to multiple parties through a central exchange. This system of exchanges evolved over time, becoming more automated and more efficient, making telephone communication more accessible to the general public.

The spread of the telephone had a democratizing effect. Rural areas, previously isolated, began to link with urban centers, integrating communities and allowing for the more equitable distribution of information and resources. This shift enhanced social cohesion and allowed for the faster dissemination of news, cultural practices, and innovations. Families separated by vast distances could stay in touch more intimately and immediately, shrinking emotional distances alongside geographical ones.

As the technology advanced, so did its applications. The invention of the switchboard and the subsequent automation of telephone exchanges facilitated the rapid expansion of telephone networks. These networks formed the backbone of modern communication, providing the infrastructure that would eventually support the internet and mobile technologies. The establishment of long-distance lines and the development of undersea cables allowed for international calls, further shrinking the globe.

The social impact of the telephone was also profound. It altered individual lifestyles, making it easier to maintain relationships over distances. Emergency services could be summoned with a quick call, saving countless lives. Business operations could be coordinated across different locations, creating new efficiencies and opportunities. The telephone laid the groundwork for a new kind of economy, one that was faster, more dynamic, and more interconnected.

The telegraph and telephone, when viewed together, symbolize the human endeavor to overcome the limitations of distance and time. They represent not just technological advancements but profound shifts in how societies function and evolve. From the clicking sounds of Morse code to the familiar ring of a telephone, these inventions democratized information, brought people closer, and paved the way for an era defined by rapid and reliable communication.

What makes these innovations truly remarkable is their legacy. They set a precedent for continuous improvement and innovation in communication technologies. The principles and infrastructure established by the telegraph and telephone were instrumental in the development of future technologies, including radio, television, and the internet. Each subsequent leap in communication has built on the foundations laid by these early pioneers, continually enhancing our ability to connect and share.

Looking back, it's clear that the telegraph and telephone were more than just tools; they were catalysts for change. They broke down barriers, creating a interconnected world and fostering a global community. Their impact is still felt today, echoed in the smartphones and internet connections that have become indispensable in modern life. The march of progress that began with these simple yet revolutionary devices shows no signs of slowing down, promising even more transformative changes in the future.

Origins of Modern Communication Systems

In the grand tapestry of technological evolution, communication systems stand as pivotal threads weaving societies closer, defying distances, and forming the backbone of our interconnected world. The journey through the origins of modern communication systems is not merely a trek through technological advancements but a fascinating story of human ingenuity and the unrelenting pursuit of connectivity. From the rudimentary signals of ancient times to the sophisticated digital networks of today, the transformation has been nothing short of revolutionary.

The embryonic stage of modern communication can be traced back to the development of the telegraph in the early 19th century. Samuel Morse and other pioneering inventors harnessed electricity to send coded messages over vast distances, evolving from optical telegraphs and beacons used in earlier centuries. The telegraph system was nothing less than groundbreaking; it converted information into electric signals and then back into readable text. This leap turned continents into neighbors and made near-instantaneous messaging a reality.

But it wasn't just hardware that mattered. The Morse code system revolutionized how messages were encoded and transmitted. It was a universal language of dots and dashes that transcended spoken language barriers and unified global communication in an unprecedented way. The establishment of undersea telegraph cables connecting continents marked a colossal stride in international connectivity, laying the foundation for a globally networked society.

Next in our journey was Alexander Graham Bell's invention of the telephone in 1876. The telephone transformed the potential of communication from text-based messages to voice-based interactions, making it even more personal and immediate. For the first time, people could hear each other's voices across miles, lending an intimacy to conversations that the telegraph could never achieve. This innovation shifted the paradigm from public messaging systems to more private and personal communication channels, compelling society to re-evaluate notions of privacy and connectivity.

The telephone network's expansion was rapid and far-reaching. Telephone poles and wires began crisscrossing cities and countrysides, bringing voice communication into homes and businesses around the world. This seamless voice connectivity laid the groundwork for the collaborative and communicative modern business landscapes we know today. It revolutionized industries, facilitating real-time decision-making and strategic discussions, which previously were constrained by the delays of written correspondence.

As these early communication systems evolved, they relied heavily on physical wiring. Enter radio communication—an innovation that fundamentally altered the landscape by removing the necessity for wired connections. Guglielmo Marconi's successful demonstration of wireless telegraphy in the late 19th century unlocked a new realm of possibilities. Radio waves could carry signals over both short and long distances without the constraints of physical infrastructure, opening communication channels to ships at sea and remote locations.

Radio communication also found its way into the homes of millions through the advent of broadcast radio in the early 20th century. The concept of broadcasting allowed a single signal to be transmitted to many receivers simultaneously, revolutionizing mass communication. News, entertainment, and educational content could now be disseminated broadly and rapidly, democratizing information and setting the stage for the mass media age. The socio-political impact was profound, as governments and organizations began using radio to galvanize public opinion and coordinate large-scale efforts in ways previously unimaginable.

World War II further accelerated advancements in communication technology. The necessity for secure, rapid, and reliable communication channels spurred innovations in coding, encryption, and signal processing. The groundwork laid during these years was instrumental in developing more advanced communication systems, facilitating the leap into the digital age.

Post-war advancements saw the birth of the transistor and subsequently, the era of digital communication. The invention of transistors allowed for the miniaturization and increased efficiency of electronic components. This technological progression was a linchpin for modern computing and digital communication systems. The shift from analog to digital wasn't just a shift in technology; it was a paradigm change in how information was processed, stored, and transmitted.

By the mid-20th century, satellite communication emerged, unlocking new capabilities in global connectivity. The launch of the first communication satellite, Telstar, in 1962, demonstrated the potential for relaying telephone and television signals across continents without the need for extensive cable networks. Satellites effectively shrank the world, making global live broadcasts possible and paving the way for the immense global telecommunication networks that support modern society.

The late 20th century witnessed the convergence of computing and communication technologies, giving birth to the internet. What started as a modest network for academic and military use, ARPANET, quickly grew into a global digital communication network. The development of packet-switching technology and TCP/IP protocols laid the sturdy foundation for the internet, allowing diverse networks to interconnect and communicate seamlessly. The promise of instant, global communication gradually became a reality, revolutionizing every aspect of human life—from how we conduct business to how we socialize.

By the 1990s, personal computers became household staples and the World Wide Web, spearheaded by Tim Berners-Lee, added a rich, hyperlinked dimension to the internet. Websites became the new information hubs, and email transformed personal and professional correspondence. The innovation of search engines, social media platforms, and eventually smartphones further amplified the utility and reach of the internet, embedding communication deeply into the fabric of daily life.

Each step in this evolutionary journey—from the telegraph and telephone to radio waves, satellites, and the digital revolution—represents the relentless drive of human innovation to bridge distances and foster connections. In understanding the origins of modern communication systems, we're not just tracing technological milestones; we're exploring the very essence of what it means to be human in an ever-connected world. Through these advancements, society has experienced profound changes in the way we interact, share knowledge, and build relationships. These innovations have not only connected us but have also empowered us, creating new avenues for collaboration, creativity, and progress.

Modern communication systems continue to evolve, now encompassing emerging technologies like 5G networks, quantum communication, and the Internet of Things (IoT). Each of these burgeoning fields promises to push the boundaries of connectivity further, heralding new revolutionary changes. The story of communication advancements is ongoing, ever pushing the envelope of what's possible, and continually reshaping our world in ways we are only beginning to imagine. The relentless march of progress in communication technology mirrors humanity's unyielding quest for ever-greater understanding, connection, and collaboration.

CHAPTER 12:
THE AGE OF ELECTRICITY

The Age of Electricity marks a transformative period in human history, illuminating and powering the modern world in ways once thought impossible. It began with pioneering figures like Edison and Tesla, whose breakthroughs in electrical engineering laid the groundwork for an electrified future. With the flick of a switch, homes, streets, and factories scintillated to life, reshaping daily existence and industrial operations. This era catalyzed an explosion of innovation, from the incandescent light bulb to the electric motor, revolutionizing how we live, work, and communicate. Electricity not only fueled the Second Industrial Revolution but also bridged the gap between invention and accessibility, making technological marvels a part of everyday life and setting the stage for future advancements.

Pioneers of Electrical Engineering

The Age of Electricity wouldn't have sparked without the remarkable minds that pioneered electrical engineering. Their discoveries not only illuminated our homes but also laid the groundwork for the modern technological landscape. It's here that we dive into the lives and contributions of the visionaries whose ideas transformed the world.

Michael Faraday, often revered as the father of electrical engineering, initially worked as a bookbinder. His curiosity and passion for knowledge led him to attend lectures by the eminent scientist Humphry Davy. Faraday's relentless enthusiasm caught Davy's eye, securing him a position at the Royal Institution. There, Faraday made groundbreaking discoveries in electromagnetism and electrochemistry. His laws of electrolysis and induction are cornerstones of modern electrical engineering, encapsulating phenomena that were once mysteries. Despite his lack of formal education, Faraday's ingenuity illuminated the path for countless engineers and scientists.

Another cornerstone in the field, James Clerk Maxwell, translated Faraday's concepts into a set of mathematical equations. Maxwell's equations, as they're known today, form the foundational framework for understanding electromagnetism. His theoretical insights enabled the quantification and manipulation of electrical phenomena, bridging the gap between abstract theory and practical application. Maxwell's work didn't just advance electrical engineering; it also spurred on developments in physics and optics.

Thomas Edison, although more famous as an inventor than a theorist, played an indispensable role in popularizing the use of electricity. Edison's relentless experimentation led to the development of the incandescent light bulb, a safer, longer-lasting alternative to gas lamps. But Edison's contributions went beyond this single invention. His establishment of electric power stations and distribution systems revolutionized urban living, effectively turning night into day. Edison's dogged determination and business acumen brought electricity to the masses, making it an integral part of daily life.

Alongside Edison, we find Nikola Tesla—a name synonymous with innovation and futuristic thinking. Originally working under Edison, Tesla's vision diverged, leading him to explore alternating current (AC) as opposed to Edison's direct current (DC). Tesla believed AC had the potential for more efficient transmission over long distances. His work in alternating current eventually led to the development of the AC motor and transformer, technologies still in use today. Tesla's ideas were well ahead of his time, laying the groundwork for wireless communication and even early visions of modern computing.

Another trailblazer, Heinrich Hertz, was the first to conclusively prove the existence of electromagnetic waves, the very basis of radio technology. Through meticulous experimentation, Hertz demonstrated that these waves could be transmitted and received, validating Maxwell's theoretical predictions. Though his work focused more on proving concepts than practical applications, Hertz's discoveries paved the way for a whole new world of wireless communication.

Guglielmo Marconi took Hertz's theories and transformed them into tangible technology. Often hailed as the "father of radio," Marconi's work in wireless telegraphy made it possible to send signals across vast distances without the need for physical connections. This revolutionary advancement connected continents, catalyzing the growth of global communication networks. Marconi's inventive spirit and commercial success showed how theoretical concepts could be transformed into practical tools that altered the fabric of society.

In the early 20th century, Charles Proteus Steinmetz made lasting contributions to the understanding of alternating current systems. His work on transient electrical phenomena and hysteresis fundamentally changed how engineers approached electrical engineering problems. Steinmetz's ability to combine theoretical knowledge with practical applications allowed for greater efficiency and stability in electrical systems, many of which serve as the backbone of modern infrastructure.

The contributions of these pioneers were not confined to theoretical discoveries or individual inventions; they forged new industries, reshaped economies, and built the backbone of modern technology. The symbiotic relationship between their inventive minds and practical applications illustrates the transformative power of electrical engineering. Each innovation built upon the last, demonstrating how human curiosity and determination can light up entire worlds.

Undoubtedly, the collective work of these pioneers and countless others has sown the seeds for today's extraordinary technological advancements. Their legacy continues to inspire engineers, scientists, and dreamers, showing that even the most formidable challenges can be overcome with ingenuity and perseverance. The Age of Electricity is an enduring testament to the extraordinary impact a few determined individuals can have on the course of history.

Electrifying Daily Life

Imagine a world shrouded in darkness once the sun dipped below the horizon, where evening activities were limited by the flickering and often unreliable glow of candles and oil lamps. This was the reality of life before the advent of electricity. The shift from such a world to one illuminated by electric lights was nothing short of transformative. Electric lighting turned night into day, and for the first time, people could extend their activities and productivity well into the night. It was a change that not only brightened homes and cities but also catalyzed further innovations and fundamentally altered the rhythm of daily life.

Electric lights were more than just a convenience; they revolutionized industries and transformed entertainment. Theatres and concert halls, once constrained by the limitations of gas and candlelight, came alive with brighter and safer electric lighting. This technological leap greatly enhanced the audience experience and increased the venues' capacity for events, pushing the boundaries of what was previously possible.

Factories, too, found new life with electric lighting and power. Production lines became more efficient as work could continue reliably regardless of natural light. This led to increased productivity and, ultimately, economic growth. The relentless hum of machinery and the bright, steady light of the factory floor symbolize the era's relentless drive towards progress and efficiency. These advances also coincided with the development of electric streetlights, which began to illuminate urban landscapes, making cities safer and more vibrant after dark.

The impact of electricity on domestic life was equally profound. The introduction of household appliances such as electric irons, vacuum cleaners, and washing machines began to ease the burden of daily chores. Tasks that once demanded significant physical effort and time could now be accomplished with mere button presses. This had a particularly liberating effect on women, who traditionally bore the brunt of household work, enabling them to engage in activities outside the home, including education and professional careers.

Communication, too, was revolutionized. The telegraph had already started shrinking distances, but the advent of electrically powered devices like the telephone allowed for instantaneous voice communication over long distances. This breakthrough not only spread personal connections but also facilitated business operations, helping to knit the fabric of society even tighter.

It's hard to overstate the excitement and sense of possibility that characterized the early days of household electricity. Imagine, for a moment, the sheer wonder of flipping a switch and having a room flood with light. No smoke, no delay, just the instantaneous magic of illumination. The psychological impact of this newfound control over one's environment cannot be underestimated. It fostered an era of innovation and encouraged an attitude that almost anything was possible with the right invention.

Beyond lighting and basic appliances, electricity paved the way for the myriad devices that define modern life. Radios brought news, music, and entertainment into homes, shrinking the world further by connecting listeners to events and cultures far beyond their immediate surroundings. Refrigerators changed food storage practices, leading to fresher diets and improved public health. These electrical marvels became household staples, marking progress with every new iteration.

Electricity also became a fundamental part of social and cultural life. Electric trams and subways redefined urban transport, making it easier for people to commute, thereby transforming cities and enabling suburban expansion. The electric lift, or elevator, allowed for the construction of skyscrapers, effectively reshaping city skylines and enabling more efficient use of vertical space in densely populated areas.

Yet, the journey wasn't without its challenges. The widespread distribution of electricity required monumental infrastructure projects. Power plants had to be built, and an extensive network of cables and transformers was needed to deliver electricity to homes, businesses, and factories. These projects were embodiments of human ingenuity and determination, showcasing our ability to reshape the environment to meet our needs.

The efforts to harness and distribute electrical power demanded robust industrial coordination. Companies like Thomas Edison's General Electric and Nikola Tesla's Westinghouse Electric played pivotal roles. Their fierce competition, often referred to as the "War of Currents," was a defining chapter in the history of electricity. Edison's direct current (DC) and Tesla's alternating current (AC) systems vied for dominance, each with its advantages and disadvantages. Ultimately, the efficiency and easier long-distance transmission of AC won out, but the rivalry spurred rapid advancements and innovations in electrical engineering.

Public acceptance of this new technology wasn't instant. There were fears and misconceptions surrounding the safety of electricity. Educational campaigns and public demonstrations were essential to overcoming these barriers. Once the safety and reliability of electrical systems were established, trust grew, and electricity became a staple of modern living.

The electrification of rural areas presented another set of challenges. In many regions, especially in vast countries like the United States, extending electrical grids to remote areas was a logistical and economic hurdle. Initiatives like the Rural Electrification Administration (REA) of the 1930s aimed to address these disparities, ensuring that the benefits of electricity reached even the most isolated communities. Electrification of rural areas transformed agricultural practices, making them more efficient and less labor-intensive, thus contributing to the overall growth and modernization of these regions.

Moreover, electricity laid the groundwork for future technological innovations that would take the world by storm. The development of electric-powered computers, for instance, set the stage for the digital revolution. Every device that forms part of our daily lives today, from smartphones to electric cars, owes its existence to the foundational breakthroughs in electrical engineering.

The transition to a world powered by electricity also had environmental and social implications. Initial power generation relied heavily on fossil fuels, which introduced concerns about pollution and resource depletion. These early concerns paved the way for contemporary discussions and innovations in renewable energy sources, seeking to balance the benefits of electricity with sustainable practices.

It's fascinating to see how electricity went from a marvel of scientific curiosity to a fundamental component of modern life. The journey of harnessing electrical power wasn't just a series of technical achievements; it was a cascade of transformations that touched every aspect of society. It catalyzed innovations that altered the dynamics of industry, reshaped urban and rural landscapes, and redefined what was possible in personal and professional lives.

As we look to the future, the legacy of the age of electricity serves as both a foundation and an inspiration. The initial challenges and triumphs of electrification remind us that progress often involves overcoming substantial obstacles. Yet, the rewards of persistence and innovation can be profound, illuminating not only our homes and cities but also the path to an ever-brighter future.

CHAPTER 13:
TRANSPORTATION TRANSFORMED

Transportation, a vital artery of human progress, underwent a remarkable metamorphosis in the 20th century, reshaping how we live, work, and connect. The advent of automobiles began to render distances irrelevant, creating unprecedented personal mobility and unlocking vast economic possibilities. Meanwhile, the skies became not just a realm of dreams but a tangible frontier, with the first powered flights by the Wright brothers heralding a new era of aviation that would eventually shrink the world. Railways and ships, the lifeblood of earlier centuries, adapted and evolved, integrating advanced technologies and becoming even more pivotal in global trade and mass transit. These breakthroughs collectively created a world more intricately linked than ever before, fundamentally transforming societies and economies by collapsing geographical barriers and fostering a new era of global interconnectedness.

Automobiles and Airplanes

Our journey through the history of transportation takes a thrilling leap with the advent of automobiles and airplanes, the twin pillars of modern mobility. These innovations changed the fabric of our daily lives, shaping not just how we travel, but also how we perceive distance and time. They marked a decisive shift from the sedentary constraints of earlier epochs to an era where speed and efficiency reign supreme.

Automobiles, in particular, revolutionized personal mobility. Prior to their arrival, most people relied on horses, carriages, or simply walking to get from one place to another. While railways offered speed and distance, they were bound by tracks and timetables. The automobile, however, offered a sense of freedom that was previously unimaginable. Early models, like Henry Ford's Model T, were not just engineering marvels but also social equalizers. They democratized transportation, making it accessible to the masses and reshaping urban landscapes.

The impact of the automobile extends beyond personal convenience. It transformed industries, spawning a vast network of roads, highways, and infrastructure projects. Gas stations, repair shops, motels, and diners sprouted up along newly paved roads, giving rise to the phenomenon of car culture. Moreover, the automobile catalyzed economic growth, prompting advances in manufacturing techniques and creating millions of jobs. The assembly line production pioneered by Ford didn't just revolutionize automobile manufacturing; it set a new standard for industrial efficiency across multiple sectors.

As we trace the development of the automobile, it's essential to consider the technological advancements that propelled its evolution. From the inception of the internal combustion engine to advancements in design and safety features, each step forward was built on the ingenuity and perseverance of inventors who dared to dream. Innovations like automatic transmissions, power steering, and anti-lock brakes have since become standard, continually pushing the envelope of what is possible on four wheels.

In tandem with the rise of the automobile, the skies opened up to human exploration and travel with the invention of the airplane. The Wright brothers' first successful flight in 1903 was a defining moment not just for transportation but for human history. It symbolized mankind's relentless quest to conquer the skies and defy the earthbound limitations that had previously governed our existence.

The initial years of aviation were filled with experimentation and daring. Enthusiasts and engineers alike continuously pushed the boundaries, resulting in aircraft that could fly higher, faster, and farther. World War I served as a crucible for aviation technology, as airplanes shifted from reconnaissance roles to active combat and transport missions. By World War II, the advances were staggering, with bombers and fighters demonstrating unprecedented capabilities.

The post-war era saw aviation evolve from a military necessity to a cornerstone of global commerce and travel. Commercial aviation took center stage, with iconic aircraft like the Boeing 707 and the Douglas DC-3 making air travel accessible to the general public. Airports sprung up around the world, and the concept of the global village became more tangible as people could cross continents in mere hours.

The development of jet engines marked a paradigm shift. Airplanes became faster, more reliable, and more efficient, leading to the age of mass air travel. Today, the aviation industry is a testament to human ingenuity, hosting some of the most sophisticated technologies ever designed. From computer-aided navigation systems to advanced aerodynamics, modern airplanes epitomize the extraordinary progress we've made.

Yet, the story of automobiles and airplanes isn't just about technological achievements. It's also about how these innovations have woven themselves into the very fabric of society. Cars and planes have not only transformed our physical landscape but have also impacted our cultural and social frameworks. They've influenced everything from urban planning and international trade to family vacations and daily commutes.

Consider the automobile's role in suburbanization. As cars became more prevalent, people were no longer confined to living close to where they worked. Suburbs sprang up, offering a blend of urban and rural living, and profoundly altered the social and economic dynamics of cities. This shift enabled a new way of life, characterized by commuting and a growing emphasis on home ownership away from city centers.

Similarly, airplanes have made the world a smaller place. The ability to travel across oceans in a matter of hours has broadened our horizons, fostering cultural exchange and global interconnectedness. Business deals that once took months to negotiate can now be finalized in a mere afternoon across continents. Families separated by great distances can reunite with relative ease, making the world feel more intimate despite its vastness.

The legacy of these modes of transportation also extends into the realm of environmental impact. While they've brought unparalleled convenience and connectivity, they've also posed significant challenges. The automobile's dependence on fossil fuels has contributed to pollution and climate change, prompting urgent discussions and innovations around sustainable energy and electric vehicles. Similarly, the aviation industry's environmental footprint is significant, sparking ongoing research into greener technologies and alternative fuels.

Looking ahead, both automobiles and airplanes are on the cusp of further transformations. Electric and autonomous vehicles are poised to redefine personal transportation, offering possibilities for safer, more efficient, and environmentally friendly travel. Similarly, the aviation industry is exploring electric planes, super-efficient jet engines, and even commercial space travel.

The journey of automobiles and airplanes from their rudimentary beginnings to their current state is a testament to human creativity and resilience. These innovations have not only changed how we move but have also affected every facet of our lives, from the economy to the environment, and from our daily routines to our collective consciousness. As we continue to innovate, the very essence of mobility will keep evolving, carrying us forward into uncharted territories.

In closing, the story of automobiles and airplanes encapsulates the spirit of human aspiration. They remind us that with curiosity, perseverance, and the courage to dream, we can transcend our limitations and achieve the seemingly impossible. This spirit will undoubtedly continue to drive us as we navigate the future, ever eager to explore new frontiers and redefine what's possible in the realm of transportation and beyond.

Railways and Ships

The advent of railways and ships marked a dramatic shift in the way people moved goods and traveled, forever reshaping our global landscape. Railways catalyzed the massive industrial and societal transformations of the 19th century, while ships bridged continents and oceans, making distant lands more accessible than ever before. This blend of land and maritime advancements allowed for unprecedented scale in trade, communication, and cultural exchange.

Starting with railways, the 19th century saw their rise as a revolutionary force. The first commercial railway, the Stockton and Darlington Railway, opened in England in 1825. It was a modest beginning, largely intended for coal transport. However, the success of such operations didn't take long to catch the public's imagination. Soon, railways became the arteries of nation-states, pumping economic lifeblood through burgeoning cities.

As railways expanded, they democratized travel. What was once the purview of the wealthy became accessible to the masses. The Stanforth family could visit the Hastings relatives, all within a day's journey. A farmer could transport fresh produce to urban markets overnight. The efficiency and speed of trains accelerated industrial growth, fostering interconnected economies that laid the groundwork for globalization.

The technological innovations didn't stop there. The subsequent standardization of gauge sizes and the development of more powerful locomotives meant longer and faster routes. The iconic transcontinental railroads, such as America's Transcontinental Railroad completed in 1869, were nothing short of engineering marvels. They spanned thousands of miles of challenging terrain, stitching together regions that had previously felt like disparate worlds.

Parallel to the railway revolution, shipbuilding was also sailing into a new era. The Age of Sail, marked by majestic clipper ships and tall masts, gave way to the Age of Steam. Steam propulsion transformed oceanic voyages, obliterating the whims of wind and weather that had long dictated maritime travel. The steamship SS Great Western, launched in 1838, could cross the Atlantic in just 15 days, a remarkable feat for its time.

While railways knitted the borders of nations, ships forged global networks. The opening of the Suez Canal in 1869 and the Panama Canal in 1914 were milestones that redefined trade routes. Maritime pathways shortened, linking East and West in an intricate web of commerce. Ships became the backbone of international trade, ferrying goods, people, and cultures across vast oceans.

Naval architecture continued to evolve, with steel replacing wood and hydraulic cranes transforming port operations. Shipyards buzzed with activity, producing vessels of increasing size and complexity. The dreadnought battleships, emerging in the early 20th century, epitomized technological prowess and power projection on a global scale.

Yet, the importance of railways and ships extended beyond mere physical transportation; they became symbols of progress and modernization. They were conduits for ideas, spreading innovations as widely as they spread goods. Newspapers, books, and mail could travel farther and faster, influencing societies far removed from their points of origin. The rhythms of life synchronized with the timetables of trains and the schedules of steamships.

Of course, such rapid industrial and transportation advancements were not without their challenges and downsides. The construction of railways disrupted traditional lands and communities. Indigenous populations were often displaced or coerced into labor. Similarly, the shipping industry's drive for profit contributed to environmental degradation and the exploitation of workers. The global appetite for rubber, cotton, and other raw materials fed these vast networks, often at the expense of colonized regions.

Despite these challenges, the overall impact of railways and ships has been profoundly transformative. They have enabled the large-scale movement of people, leading to urbanization and the development of multicultural cities. The shipping industry, meanwhile, laid the logistical foundations for modern supply chains. Every container ship docking at a port today owes its existence to the advancements made during this transformative period.

It's also worth noting the remarkable human stories intertwined with these technological marvels. The engineers who designed intricate railways through mountains, the sailors who bravely navigated uncharted waters, and the ordinary workers who laid down endless tracks and riveted steel hulls — their collective efforts have created a legacy that continues to impact our lives.

Modern transportation methods, whether high-speed trains or container ships, still carry the DNA of these pioneering eras. Magnetic levitation trains, cutting-edge shipping logistics, and completely autonomous vessels are today's beneficiaries of centuries-long innovation. Their origins in the railways and ships of the past serve as a testament to human ingenuity. As we glance back at the tracks and sea lanes etched into our history, it becomes evident that our present, and indeed our future, is built on the enduring foundations laid down by these early transportation titans.

- **Efficiency:** Dramatically reduced travel times and costs.

- **Accessibility:** Made travel affordable for a broader population.

- **Global Trade:** Expanded markets and facilitated international commerce.

- **Economic Growth:** Spurred industrial expansion and urbanization.

In sum, railways and ships not only transformed transportation—they transformed the world itself. The iron rails and steaming hulls bridged gaps between towns, nations, and continents, accelerating the pace of human progress like never before. It's a narrative rich with both remarkable achievements and significant challenges, each leaving an indelible mark on the tapestry of human history.

CHAPTER 14:
THE BIRTH OF MODERN MEDICINE

The emergence of modern medicine marked a profound shift in humanity's quest to combat disease and extend life. Building upon centuries of rudimentary practices, this transformative era brought groundbreaking advances like vaccination and germ theory, fundamentally altering our understanding of illness and health. The application of these scientific discoveries led to significant strides in medical treatment and public health. Innovations such as anesthesia and antiseptic techniques revolutionized surgery, reducing mortality rates and making procedures safer and more effective. This period laid the foundational principles for contemporary medical science, propelling us out of the darkness and into an age where empirical evidence and rigorous research became the cornerstones of healthcare. The birth of modern medicine not only extended human lifespan but also dramatically improved the quality of life, underscoring the incredible potential of science and technology to drive social progress.

Vaccination and Germ Theory

The dawn of modern medicine wasn't marked by a single invention but rather a constellation of discoveries and innovations that transformed our understanding of health and disease. Central to this transformation were the concepts of vaccination and germ theory, revolutionary ideas that turned the tide against many of humanity's most deadly afflictions. While they emerged separately, these two pillars of modern medical science are inextricably linked in their profound impact.

In the late 18th century, smallpox was a devastating disease that claimed millions of lives. The breakthrough in vaccination came from the unlikely interplay of observation and experiment. In 1796, Edward Jenner, an English physician, noticed that milkmaids who had contracted cowpox, a less severe disease, seemed immune to smallpox. Inspired by this observation, he inoculated an eight-year-old boy with material from a cowpox sore and later exposed him to smallpox. Remarkably, the boy did not develop the disease. Jenner's work paved the way for the concept of vaccination, derived from "vacca," the Latin word for cow.

Vaccination faced initial skepticism. Jenner's methods were unorthodox, and the idea of introducing material from one disease to prevent another was seen as radical. However, as further experiments confirmed his results, the medical community began to embrace vaccination. By the 19th century, it had become a widespread practice, saving countless lives and becoming a cornerstone of public health.

The story of germ theory follows a somewhat parallel but distinct path. For centuries, the prevailing belief was that diseases were caused by "miasma," or bad air. This theory persisted despite mounting evidence to the contrary. It wasn't until the mid-19th century that a few pioneering scientists began to challenge the status quo, leading to the germ theory of disease, which posited that microorganisms, rather than airy vapors, were the culprits behind many illnesses.

Louis Pasteur, a French chemist, was at the forefront of this revolution. Pasteur's meticulous experiments with fermentation demonstrated that microorganisms were responsible for both the process of fermentation and the spoilage of food. This insight had profound implications for human health. Pasteur went on to develop vaccines for rabies and anthrax, further bolstering the acceptance of germ theory.

His work not only saved lives but also laid the foundation for microbiology as a scientific discipline.

Simultaneously, Robert Koch, a German physician, made significant strides in proving the germ theory through his work on tuberculosis. Koch's postulates, a series of criteria designed to establish a causal relationship between a microbe and a disease, became essential tools for identifying and studying pathogens. Koch's contributions were immense, earning him the Nobel Prize in Physiology or Medicine in 1905 and solidifying his place in medical history.

Despite their revolutionary nature, neither vaccination nor germ theory would have had the impact they did without broader social and technological contexts. The Industrial Revolution, for instance, brought about urbanization and densely populated cities, which were hotbeds for disease. These conditions necessitated rapid advancements in public health initiatives, making the implementation of vaccination programs and germ theory-based sanitation measures all the more crucial.

One of the most notable examples of this interplay was the cholera outbreaks of the 19th century. When London was gripped by cholera, Dr. John Snow, a staunch supporter of germ theory, traced the outbreak to a contaminated water pump. By removing the pump handle, he curtailed the spread of the disease, demonstrating the practical applications of germ theory in public health.

The acceptance of vaccination and germ theory also owes much to the burgeoning field of medical education. Schools and universities began to incorporate these concepts into their curricula, training generations of physicians and healthcare professionals. This formal education ensured that the benefits of these innovations were widely understood and applied, making them integral to modern medical practice.

In addition to the immediate impact on disease control, these advancements have had long-lasting effects on medical research and practice. The principles of vaccination underpin modern immunology, while germ theory informs our understanding of infectious diseases, leading to the development of antibiotics, sterilization processes, and advanced diagnostic techniques. Today, the legacy of these breakthroughs can be seen in the robust response mechanisms to new diseases and pandemics, highlighting their enduring relevance.

Moreover, the societal implications of these discoveries extend beyond the realm of medicine. The concept of herd immunity, which relies on a significant portion of the population being immunized to protect those who cannot be vaccinated, underscores a communal approach to health. This idea has fostered a sense of collective responsibility and interdependence, serving as a foundation for public health policies globally.

In conclusion, the birth of modern medicine, marked by the advent of vaccination and germ theory, represents a turning point in human history. These innovations not only revolutionized the way we understand and combat diseases but also set the stage for future medical advancements. Through the persistent efforts of pioneers like Jenner, Pasteur, and Koch, the foundations of public health were laid, transforming societies and saving millions of lives. As we continue to face new health challenges, the lessons learned from these early triumphs remain as relevant as ever, guiding us toward a healthier, more resilient future.

Surgical Advances

The dawn of modern surgery is one of the most transformative chapters in the history of medicine. While ancient civilizations like Egypt and Greece practiced rudimentary surgical techniques, the foundational advances that define modern surgery began to take shape in the

19th and 20th centuries. These innovations not only reshaped medical practice but also played a crucial role in extending human life expectancy and improving the quality of countless lives.

One of the first significant milestones in surgical history was the development of anesthesia. Before its invention, surgery was an agonizing experience, often performed only when absolutely necessary. The introduction of ether in the mid-19th century by William T.G. Morton and the subsequent use of chloroform by Sir James Young Simpson revolutionized the field. Patients could now undergo procedures without experiencing unbearable pain, allowing surgeons to perform more intricate and life-saving operations.

In tandem with anesthesia, the advent of antiseptic techniques marked another monumental leap. Ignaz Semmelweis and Joseph Lister championed the use of handwashing and sterilization methods to combat the spread of infections during surgery. Lister's use of carbolic acid to sterilize surgical instruments and clean wounds significantly reduced post-operative infections and mortality rates. This was a game-changer, providing a much cleaner and safer environment for patients and drastically enhancing surgical outcomes.

As antiseptic methods evolved, so did the tools at a surgeon's disposal. The 19th and early 20th centuries saw incredible advancements in surgical instruments. Scalpels, forceps, and clamps became more sophisticated, allowing for greater precision and control. The development of surgical sutures, such as catgut and silk, improved wound closure techniques and promoted better healing. Enhanced instrumentation turned surgery from a crude, often dangerous endeavor into a refined and effective medical practice.

The rise of diagnostic imaging also played a pivotal role in the evolution of modern surgery. Wilhelm Conrad Roentgen's discovery of X-rays in 1895 offered a non-invasive way to peer inside the human body. Surgeons could now plan their procedures with greater accuracy, identifying the precise location and extent of abnormalities. This led to better surgical planning and reduced complications, further cementing the place of surgery in the modern medical landscape.

Moving into the mid-20th century, the scope of surgical possibilities expanded dramatically with the introduction of antibiotics. Alexander Fleming's serendipitous discovery of penicillin in 1928 ushered in a new era in which bacterial infections, once the bane of surgical procedures, could be effectively treated. The widespread use of antibiotics drastically reduced post-surgical infections, making previously high-risk procedures much safer.

As the 20th century progressed, so too did the intricacy and scope of surgical innovations. One notable area of advancement was cardiovascular surgery. The first successful open-heart surgery, performed by Dr. John Gibbon using a heart-lung machine in 1953, opened the door to a plethora of life-saving cardiac procedures, including coronary artery bypass and valve replacements. These breakthroughs have saved millions of lives and continue to evolve, with minimally invasive techniques reducing recovery times and complications.

Transplantation surgery is another field that saw monumental advances. The first successful kidney transplant in 1954 by Dr. Joseph Murray set the stage for future organ transplants. Heart, liver, lung, and pancreas transplants soon followed, offering hope to patients with previously incurable diseases. Advances in immunosuppressive drugs, particularly the development of cyclosporine in the 1980s, significantly improved transplant success rates by preventing organ rejection.

Minimally invasive surgery, which emerged in the latter half of the 20th century, represents one of the most exciting and ongoing areas of surgical innovation. Techniques such as laparoscopy, where surgeons use small incisions and a camera to perform procedures, have transformed patient care. These methods result in less pain, quicker recovery times, and lower risk of complications compared to traditional open surgeries. Robotics further augments these capabilities, allowing for even greater precision and the ability to perform complex surgeries with enhanced accuracy.

Robotic-assisted surgery, brought to the forefront by systems like the Da Vinci Surgical System, has revolutionized how surgeons approach patient care. These robots provide unparalleled dexterity and vision, allowing surgeons to perform intricate procedures through small incisions. The technology continues to evolve, with advancements in robotic systems making surgeries safer and more accessible worldwide.

Innovation in surgical techniques isn't just confined to the tools and methods used but also extends to the very environments where surgeries are performed. The advent of specialized surgical suites and sterile operating rooms has created optimal conditions for surgery, reducing the risk of infections and complications. This controlled environment, coupled with the use of advanced monitoring systems, ensures that surgeons can provide the best care possible.

But all these technological advances would be meaningless without the dedication and skills of the surgeons themselves. Advances in medical education have played a crucial role in improving surgical outcomes. Modern training programs emphasize rigorous academic grounding, hands-on experience, and continuous learning. The advent of surgical simulators and virtual reality training tools has made it possible for surgeons to practice complex procedures repeatedly, honing their skills to perfection before they ever operate on a live patient.

While the journey of surgical advances is remarkable when viewed through the lens of history, it's essential to acknowledge that it is an ongoing journey. Researchers and pioneers in the field continually push the boundaries, exploring new frontiers like regenerative medicine and bioengineering. These cutting-edge areas hold the promise of even more incredible advancements, such as growing organs in the lab, personalized surgical interventions tailored to an individual's genetic makeup, and advanced prosthetics that integrate seamlessly with the human body.

The story of surgical advances is not just one of technological triumph but also of human resilience and ingenuity. From the struggles of early pioneers who dared to imagine a world where surgery could be safe and effective, to the current era of high-tech precision and minimally invasive procedures, each step forward has been driven by a determination to alleviate suffering and prolong life. This relentless pursuit of excellence and innovation continues to inspire and shape the future of medicine.

As we look forward to the future, it's clear that surgical advances will continue to play a vital role in the ongoing evolution of modern medicine. With each new discovery and technological breakthrough, we can expect to see even more life-saving procedures, greater precision, and improved outcomes. The quest to perfect the art and science of surgery is a testament to human ingenuity and the unyielding desire to push the limits of what is possible.

The journey of surgical advances, rooted in the profound desire to heal and cure, remains a cornerstone of modern medicine's ever-evolving story. It's a journey that underscores the pivotal role technology plays in our lives, continually redefining our capabilities and redefining the boundaries of human health and wellness.

CHAPTER 15:
THE DIGITAL AGE

The Digital Age stands as one of the most transformative eras in human history, characterized by the birth of the computer and the dramatic revolution in information processing that followed. Emerging in the mid-20th century, the digital revolution fundamentally changed the way we think, communicate, and operate in everyday life. Computers evolved from room-sized behemoths to sleek devices that fit in the palm of our hands, democratizing access to information and enabling unprecedented levels of personal and professional productivity. The rapid development of digital technologies also laid the groundwork for subsequent leaps in interconnectedness and efficiency, including the rise of the internet and mobile technology. This chapter delves into the profound impact of digital innovation, exploring not only the technical advancements but also the societal shifts that have reshaped the fabric of our world. As we navigate through this technological landscape, we'll appreciate the pivotal breakthroughs that have catapulted us into an era where the digital realm influences nearly every aspect of our existence, paving the way for a future that feels both overwhelmingly complex and endlessly promising.

Birth of the Computer

The journey to the birth of the computer is one of human ambition intersecting with technological necessity. In the 19th century, the seeds of modern computing were sown by visionaries such as Charles Babbage and Ada Lovelace. Babbage conceptualized and began developing mechanical computing engines, most notably the Analytical Engine, which featured fundamental aspects of modern computers—such as the use of programmable cards. Lovelace, meanwhile, foresaw the potential of this machine, envisioning a future where it could perform tasks beyond simple arithmetic, like composing music or creating art.

Let's fast forward to the early 20th century. The notion of automated calculation grew in importance, especially during World War II. The need for accurate and rapid computation in cryptography and ballistic calculations accelerated research and development. Among the pivotal breakthroughs was the construction of the Electronic Numerical Integrator and Computer (ENIAC), recognized as the first general-purpose electronic computer. Completed in 1945, ENIAC was a behemoth, taking up 1,800 square feet, but it could perform complex calculations much faster than any human could.

However, ENIAC wasn't where the story truly began. It was merely the product of decades of theoretical and practical advances. Alan Turing, often hailed as the father of theoretical computer science, laid the mathematical groundwork with his concept of a universal machine capable of performing any computation that could be described algorithmically. Known as the Turing machine, this concept is fundamental to the philosophy of modern computing and continues to influence how we design software and computers alike.

As technology evolved, the invention of the transistor in 1947 by John Bardeen, Walter Brattain, and William Shockley marked a significant milestone. Transistors replaced the bulky and unreliable vacuum tubes used in earlier computers, ushering in an era of smaller, more efficient, and more reliable electronics. This moment can't be overstated; it's the dawn of modern electronics, making possible the miniaturization and performance improvements that would follow.

During the 1950s and 1960s, the field of computer science advanced rapidly. Innovations like magnetic core memory and the development of high-level programming languages such as FORTRAN and COBOL made computers increasingly versatile and powerful. This period also saw the creation of the first commercially available computers, like the UNIVAC I, which were sold to government organizations and large corporations for data processing tasks.

Next came the microprocessor, the brain of the computer, containing all the functions of a central processing unit on a single integrated circuit. In 1971, Intel released the 4004, the first commercially available microprocessor. This invention was groundbreaking. Not only did it drastically reduce the physical size of computers, but it also made them more affordable and accessible to a broader audience. Now, entire systems could be built around a single chip.

Then, something fascinating happened: computers evolved from being room-sized behemoths to personal devices. The 1970s and 1980s gave birth to the personal computer revolution. Companies like Apple and IBM brought computers into homes and small businesses, democratizing technology in a way that had never been done before. The release of the Apple II in 1977 and the IBM PC in 1981 had profound effects on personal and professional life, making the computer an integral part of daily activities.

Operating systems like MS-DOS and the advent of graphical user interfaces (GUIs) revolutionized how humans interacted with computers. Gone were the days of cryptic command-line instructions. With the introduction of the Apple Macintosh in 1984, featuring a well-designed GUI, computers began to become more user-friendly. This accessibility sparked an explosion in software development, leading to applications that catered to various needs, from business productivity to entertainment.

The evolution didn't stop there. The 1990s and early 2000s saw the rise of the internet, a global network connecting millions of computers, which transformed the computer from an isolated tool into a gateway to a vast digital world. Suddenly, the capability to share and access information across the globe was at everyone's fingertips. The development of web browsers and search engines made navigating this new digital universe possible for the average user, revolutionizing how we communicate and obtain information.

What's fascinating about the birth of the computer is its continual evolution. Moore's Law, the observation that the number of transistors on a microchip doubles approximately every two years, underlines the relentless pace of technological advancement. This exponential growth in computing power has enabled incredible innovations, from artificial intelligence to virtual reality, transforming industries and redefining human capabilities.

Moreover, today's smartphones, far more powerful than the mainframes that once filled entire rooms, epitomize the journey of the computer—from its conception to its omnipresence in our lives. But this complex tapestry of innovation isn't just about hardware. The parallel evolution of software, with languages becoming more sophisticated and operating systems more intuitive, has been equally pivotal. Each line of code, each algorithm brings us closer to unlocking new potentials.

Reflecting on the birth of the computer is to reflect on human ingenuity. It's a testament to our relentless pursuit of better solutions, driven by curiosity, necessity, and often, sheer audacity. As we continue to push the boundaries of what's possible, the foundational moments in computer history remind us of the remarkable achievements that came from daring to imagine a different future.

As we move forward in time, computers continue to evolve, becoming faster, smaller, and more integrated into our daily lives. From quantum computing to AI-driven automation, the legacy of the early pioneers lives on, continuing to shape the future trajectory of technology. Indeed, the birth of the computer isn't just a historical milestone; it's the prologue to an ongoing saga of innovation with no end in sight.

Revolutionizing Information Processing

In the grand odyssey of human progress, few innovations have redefined our world as profoundly as the revolution in information processing. Initially, the birth of the computer was seen as a marvel of engineering, a complex amalgamation of circuitry and software destined to perform calculations faster than any human ever could. But as we moved further into the digital age, this machine metamorphosed into an indispensable tool for countless facets of daily life, transitioning from profound theoretical potential to practical, ubiquitous utility.

The transformation started with early mechanical computers, like Charles Babbage's Analytical Engine, and evolved into the iconic machines of the mid-20th century, such as ENIAC and UNIVAC. These behemoths occupied entire rooms and required a small army of operators to function. Yet, beneath their veneer of complexity, they signified a new way of thinking about information—one where data could be processed at speeds and scales unimaginable to previous generations.

The breakthrough came with the transition from analog to digital. Alan Turing's fundamental concepts of algorithms and computation laid the groundwork for later advancements. By conceptualizing machines that could execute instructions stored in memory, Turing not only paved the way for modern computing but also set the stage for the revolutionary developments that would follow.

With the advent of the microprocessor in the 1970s, computers shifted from room-sized apparatuses to compact devices accessible to individuals and households. Companies like Intel and Texas Instruments played crucial roles in this leap, heralding the era of personal computing. Suddenly, information processing power was no longer confined to academic and military institutions; it was becoming democratized.

As we traversed the 1980s and 1990s, the landscape of information processing underwent another seismic shift. Software giants like Microsoft and Apple revolutionized interfaces, making computers user-friendly and increasingly integral to everyday tasks. Graphical user interfaces replaced the text-heavy command lines, simplifying interactions and broadening the appeal of computers to a global audience.

During this time, information processing was not just getting faster but also smarter. The development of relational databases, spearheaded by pioneers like Edgar F. Codd, allowed for structured storage and retrieval of vast amounts of data. This innovation led to the rise of enterprise software systems and data-driven decision-making, fundamentally altering business landscapes.

Then came the Internet, knitting together disparate computer networks into a global tapestry of information. Protocols like TCP/IP standardized communications, making it easier for systems to share data robustly and reliably. The World Wide Web, conceived by Tim Berners-Lee, took this a step further by enabling instantaneous access to information across the globe. The volume and speed at which data could be processed entered an exponential growth curve, effectively shrinking the planet into a 'global village.'

A significant milestone in this journey was the emergence of cloud computing. Service providers like Amazon Web Services, Microsoft Azure, and Google Cloud eviscerated the boundaries of physical hardware, allowing individuals and corporations to access virtually unlimited computational resources on demand. This shift unlocked new realms of possibility, from massive data storage to high-performance computing for complex problem-solving.

Parallel to these advancements, the advancements in machine learning and artificial intelligence introduced a new paradigm in information processing. Algorithms inspired by neural networks began to mimic aspects of human cognition, learning from data and making predictions with increasing accuracy. Systems like IBM's Watson showcased AI's potential by winning game shows like "Jeopardy!", solving problems in real-time that previously required human intelligence.

Modern artificial intelligence doesn't just automate tasks; it augments human capabilities, guiding decisions in fields as diverse as healthcare, finance, and logistics. Self-learning algorithms study vast quantities of data to identify patterns and provide insights that would be impossible to discern manually. Information processing has transcended mere calculations to become a cornerstone of innovation and strategic planning.

Big Data and analytics play pivotal roles in this new frontier. Today, information isn't just processed; it's mined and analyzed for actionable intelligence. Corporations harness analytic tools to sift through terabytes of data for insights that drive everything from consumer behavior predictions to optimizing supply chains. The ability to process information at such scale and granularity has radically transformed how businesses operate and compete.

Blockchain technology, often associated with cryptocurrencies, represents another transformative leap in this realm. By enabling decentralized and transparent data ledgers, blockchain has the potential to revolutionize industries by ensuring data integrity, enhancing security, and providing immutable records of transactions. This innovation further underscores the importance of sophisticated information processing in our modern economy.

Now, as quantum computing emerges on the horizon, we stand at the precipice of yet another transformation. Quantum processors leverage principles of quantum mechanics to perform computations at speeds unthinkable for classical computers. This burgeoning technology promises to solve complex problems in cryptography, materials science, and beyond, further underscoring the profound and ongoing evolution of information processing.

In education, the adaptive learning systems tailor content delivery to individual student needs, optimizing educational outcomes through the analysis of student performance data. Similarly, in healthcare, personalized medicine and diagnostics are driven by sophisticated algorithms processing voluminous patient data to offer targeted treatment plans. The implications are far-reaching, affecting every sector where data is a critical asset.

As we continue this journey, ethical considerations become increasingly paramount. Questions around data privacy, algorithmic bias, and the societal impact of automation need addressing. Striking a balance between harnessing the power of these technologies and ensuring they're used responsibly will be a critical challenge in the years to come.

The revolution in information processing is an ongoing saga, one that redefines our capabilities and reshapes our realities. It is a testament to humanity's unwavering quest for innovation and progress. As we move forward, the only certainty is that the next chapters in this story will be as unpredictable as they are transformative.

CHAPTER 16:
THE INTERNET

The Internet, from its nascent days as ARPANET to the ubiquitous World Wide Web, has revolutionized human society on an unprecedented scale. Initially a military project developed to enable robust communication, it evolved into a global network connecting billions of devices and people. The Internet has transformed how we access information, conduct business, and communicate, driving innovation in virtually every field. It has shattered geographical boundaries, brought about massive economic shifts, and enabled social movements to gain momentum at remarkable speeds. Beyond the merely technical achievements, the Internet represents a paradigm shift in human connectivity and collaboration, continuously reshaping our world and challenging us to envision new horizons for what is possible in the digital age.

From ARPANET to World Wide Web

Imagine a vast network where ideas could flow freely, connecting minds across continents in an instant. This wasn't always a reality. The journey from ARPANET to the World Wide Web epitomizes how collaboration, innovation, and ingenuity can transform society. In the early days, communicating over great distances was limited to rudimentary means, but the seeds of a connected world were sown in the turmoil of the Cold War.

ARPANET, or the Advanced Research Projects Agency Network, was not merely a product of technological ambition but a response to geopolitical tension. In the late 1960s, the United States Department of Defense sought a robust communication system that could withstand a nuclear attack. Funded by ARPA (now DARPA), researchers began constructing a decentralized network. The initial aim was simple: to link computers at different universities and research institutions, facilitating the sharing of resources. What emerged from this endeavor was far from simplistic.

On October 29, 1969, the first message was sent over ARPANET from UCLA to the Stanford Research Institute. The message, intended to be the word "LOGIN," only got as far as "LO" before the system crashed. Yet, this small step marked a monumental leap for human connectivity. By the early 1970s, ARPANET had grown, incorporating email as a standard application, which quickly became its most popular feature. This network of networks laid the groundwork for the birth of the Internet by demonstrating the feasibility and advantages of digital communication.

The innovation didn't happen in isolation. The 1970s saw the development of fundamental protocols that would define Internet communication. Transmission Control Protocol (TCP) and Internet Protocol (IP), created by Vint Cerf and Bob Kahn, established the technical foundation. These protocols ensured that data could be reliably transmitted and routed across disparate networks, enabling the seamless interchange of information. What started as a project to secure communications had ignited a paradigm-shifting evolution in technology.

Transitioning from ARPANET to what we recognize today as the Internet required more than just technical advances. It needed visionaries who could see beyond the immediate applications. By the late 1980s, Tim Berners-Lee, a British scientist, introduced a system that leveraged the existing Internet to disseminate information globally. His concept, the World Wide Web, utilized hypertext to connect documents, making navigation intuitive and user-friendly. He also developed the first browser, which transformed the process of finding and reading information online, making it accessible to a broader audience.

By 1991, the World Wide Web was available to the public. Its impact was immediate and profound. Suddenly, the labyrinth of information stored on countless disconnected computer systems became a cohesive, navigable space. The Web democratized access to knowledge, ushering in an era where information could be shared instantaneously and across geographical barriers. This is why many people often conflate the terms "Internet" and "World Wide Web," though they are technically distinct: the Internet is the infrastructure, and the Web is a service built upon it.

The 1990s were a period of explosive growth. The development of search engines, like Google, and the expansion of e-commerce platforms, such as Amazon, revolutionized how people interacted with the world. The Web became a democratizing force, allowing anyone with an Internet connection to become both a consumer and a creator of content. This interactivity fostered new communities, businesses, and forms of expression that had previously been unimaginable.

Yet, the story doesn't end there. The infrastructure has continued to evolve, driven by the relentless pace of innovation. Advances in bandwidth technologies have made streaming video, real-time communication, and responsive web applications not only possible but ubiquitous. As technologies like fiber optics and 5G networks expand, they continue to push the boundaries of what's achievable, underscoring the Internet's role as an ever-evolving artifact of human ingenuity.

The societal impacts of moving from ARPANET to the World Wide Web are varied and immense. Economies have transformed, with digital commerce creating new markets and industry models. Information has become more accessible, changing the way education, healthcare, and governance operate. Socially, the Web has redefined communities, enabling connections that transcend physical limitations, and fostering global dialogues on issues ranging from social justice to scientific collaboration.

This transformation hasn't been without its challenges. Issues surrounding data privacy, cybersecurity, and digital divides highlight the complex landscape we navigate today. Despite the Internet's promises of global unity and equality, access remains uneven, with significant portions of the world's population still unconnected. These disparities present ongoing challenges but also opportunities for future innovation and inclusion, pushing technologists and policymakers to think critically about the Internet's next chapter.

Looking back, the journey from ARPANET to the World Wide Web reflects the powerful synergy of necessity and creativity. The visionaries who laid the groundwork could not have fully envisioned how their innovations would reshape the world, yet their work has indelibly marked the trajectory of human progress. As we move forward, the Internet's story serves as a motivational beacon, reminding us that collective efforts in technology can indeed bring us closer together and unlock untold potentials.

The fabric of our interconnected world is woven from countless threads of innovation and collaboration. From the meticulous engineering of protocols to the creative genius of web pioneers, the journey from ARPANET to the World Wide Web exemplifies how transformative technology can be. It prompts us to ponder: what other frontiers await our discovery? As we continue to explore and innovate, the story of the Internet inspires us to forge ahead, ever curious about the limitless possibilities technology can reveal.

Impact on Society and Economy

It's impossible to overstate how much the Internet has transformed society and our economy. When it first began, few could have foreseen the sweeping changes it would bring. Today, it touches almost every aspect of our daily lives. From the way we communicate and shop to how we work and learn, the Internet has become a cornerstone of modern existence.

On the societal front, the Internet has dramatically altered the way we interact. Social media platforms have redefined human connection, creating virtual spaces where people can share, discuss, and even mobilize. Think of the Arab Spring or the myriad of social movements since then — all coordinated through social platforms. Information that once took days or weeks to disseminate now travels in mere seconds, breaking down barriers of distance and time.

Education has seen profound shifts as well. Online learning platforms like Khan Academy and Coursera make knowledge accessible to anyone with an internet connection. In many ways, the Internet has democratized education, offering opportunities for self-improvement and skill acquisition that were once out of reach for many.

Similarly, the workplace has undergone a seismic shift. Remote work, once a rare luxury, has become commonplace, especially in the wake of the COVID-19 pandemic. Video conferencing tools like Zoom and collaboration platforms like Slack have transformed office culture, making location almost irrelevant. This change has far-reaching implications for work-life balance, urban planning, and even environmental sustainability, as fewer people commute to work.

Economically, the impact of the Internet is just as staggering. E-commerce giants like Amazon and Alibaba have revolutionized retail, giving rise to a global marketplace that operates 24/7. Small businesses and artisans can now reach international audiences, breaking free from the constraints of brick-and-mortar establishments. The convenience of online shopping has not only changed consumer behavior but has also led to innovations in logistics and supply chain management.

Financial services have been transformed as well. Online banking and fintech startups have made financial transactions easier and more accessible. Cryptocurrencies and blockchain technology are challenging traditional monetary systems, proposing new ways to think about value, trust, and governance. Peer-to-peer lending platforms and crowdfunding websites are democratizing investment, allowing innovations to be funded without traditional gatekeepers.

Moreover, the Internet has given rise to entirely new industries. The app economy, digital marketing, online content creation—these fields didn't exist a few decades ago. The gig economy, facilitated by platforms like Uber and TaskRabbit, offers flexible work options, though not without challenges related to job security and benefits. Even in fields like journalism and entertainment, the Internet has disrupted traditional models, offering both opportunities and challenges.

While the benefits are immense, the Internet's societal and economic impacts aren't without drawbacks. Issues of privacy have become paramount, as data breaches and cyber-attacks expose sensitive personal and financial information. The debate over net neutrality underscores the tension between corporate interests and maintaining an open, fair Internet.

Misinformation and the spread of fake news pose significant challenges to public discourse and democracy. Algorithms that filter what we see online can create echo chambers, reinforcing pre-existing biases and reducing exposure to differing viewpoints. The role of technology giants in moderating content and their influence on freedom of expression are hotly debated topics.

Then there's the digital divide. Despite the Internet's global reach, access remains uneven. In many parts of the world, high-speed internet is still a luxury, perpetuating inequalities and limiting opportunities for those without reliable access. Even in developed nations, gaps in digital literacy can create significant disadvantages.

The Internet has also reshaped traditional industries. Newspapers have struggled with declining revenue as advertising dollars shift to online platforms. The music industry has had to adapt to changing models of consumption, with streaming services becoming the primary mode of delivery. These changes often require a complete rethinking of business models and strategies.

In the realm of health, telemedicine has become increasingly prevalent, breaking down geographical barriers and making healthcare more accessible. Wearables and health apps provide real-time monitoring and data collection, empowering individuals to manage their health proactively. However, these advancements also raise questions about data privacy and the security of personal health information.

The rise of Big Data analytics allows companies to make more informed decisions based on extensive data sets, offering insights previously unimaginable. This data-driven approach benefits sectors as diverse as retail, healthcare, and predictive maintenance in manufacturing. Companies that can harness the power of data are often more competitive, as they can optimize operations and better understand customer needs.

Ultimately, the Internet functions as both a unifying force and a source of division. On one hand, it connects people from all corners of the globe, fostering a sense of global community. On the other hand, it can amplify divisive rhetoric and enable harmful activities, such as cyberbullying and cybercrime. The challenge lies in navigating these complexities to maximize the benefits while mitigating the risks.

As we look to the future, it's clear that the Internet will continue to be a driving force in society and the economy. Innovations in artificial intelligence, machine learning, and the Internet of Things (IoT) promise to further integrate the Internet into our everyday lives. However, this increased integration will necessitate robust discussions around ethics, governance, and equity.

In summary, the Internet has had a profoundly transformative impact on society and the economy. It has brought remarkable benefits, democratizing access to information, revolutionizing industries, and forging new forms of human connection. Nevertheless, it also presents significant challenges that require careful consideration and thoughtful solutions. Balancing innovation with ethical considerations will be key to navigating the Internet's future trajectory.

CHAPTER 17:
MOBILE TECHNOLOGY

A s we transitioned into the late 20th century, the dawn of mobile technology began to fundamentally reconfigure our interactions, communication, and access to information. From the clunky, brick-like mobile phones of the 1980s that served primarily for voice communication, we've evolved into an era where sleek, multifunctional smartphones are indispensable. These devices are not just phones; they are powerful computing machines that fit in our pockets, blending the capabilities of computers, cameras, GPS systems, and more. The advent of mobile internet has further revolutionized our world, introducing a plethora of apps that cater to virtually every need—be it social networking, financial transactions, entertainment, or productivity. By boosting connectivity and granting instant access to vast amounts of information, mobile technology has undeniably played a pivotal role in shaping modern society, driving economic growth, and fostering innovation across countless fields. As we look forward, the ongoing advancements in mobile tech promise to continue this trajectory, offering even more sophisticated and integrated solutions to the challenges and opportunities of our ever-evolving world.

Evolution of Mobile Phones

To comprehend the profound transformation mobile phones have undergone, we must first chart their origin. The inception of mobile phones dates back to the ambitious dreams of early 20th-century visionaries who envisioned a world where telecommunication was wire-free. Motorola, a pivotal player, launched the first handheld mobile phone, the DynaTAC, in 1983. This "brick phone" was celebrated for its ability to make and receive calls without physical tethers, even though it weighed nearly two pounds and came with a hefty price tag. Yet, it was a giant leap towards a future where communication would be ubiquitous and seamless.

The early era of mobile phones was characterized by rudimentary technology and limited capabilities. The devices were large, heavy, and generally reserved for wealthy individuals or businesses that could afford their steep costs. Coverage was another significant challenge. The first-generation (1G) networks offered basic analog communication and had significant geographic limitations. But these early mobile phones laid the foundational stone for a revolution in personal communication, acting as catalysts for rapid technological advancements that shattered traditional barriers.

By the 1990s, the landscape began to shift with the introduction of the second-generation (2G) networks. Unlike their predecessors, 2G networks utilized digital encryption, enhancing both security and clarity of voice calls. Two major standards, GSM and CDMA, emerged, each competing to dominate the global market. Mobile phones began to shrink in size, shedding their unwieldy frames for sleeker, more user-friendly designs. Features like text messaging (SMS) became popular, introducing a new way for people to communicate without voice calls.

The introduction of 3G technology around the turn of the millennium further revolutionized mobile phones. These third-generation networks were the harbingers of mobile internet, enabling faster data transfer rates and expanding the scope of mobile applications. Now, mobile phones were no longer just for making calls or sending text messages. They became powerful tools capable of browsing the internet, checking emails, and even conducting business on the go. Companies like BlackBerry and Nokia thrived during this period, catering to business professionals and tech-savvy users alike.

One of the most monumental shifts came with the advent of the smartphone. In 2007, Apple unveiled the first iPhone, a device that integrated a mobile phone with a portable media player and internet browser. It wasn't just a product launch; it was a paradigm shift. The iPhone's touch interface, coupled with its app ecosystem, fundamentally changed how users interacted with their devices. The App Store, launched in 2008, created a new marketplace for software applications, enabling developers worldwide to innovate and bring new functionalities to mobile phones.

Soon, other companies followed suit. Google entered the fray with its Android operating system, a versatile and open-source alternative that quickly gained traction. Mobile phones became sleeker, faster, and significantly more powerful, each iteration pushing the envelope of what these devices could achieve. The competition between iOS and Android spurred an era of rapid innovation, with companies continually seeking to outdo each other in terms of hardware performance, software capabilities, and user experience.

The exponential growth in mobile technology during the smartphone era also heralded the rise of mobile applications, or apps, that transformed industries and daily life. From social media platforms to productivity tools, from games to health and fitness trackers, apps became an integral part of the mobile experience. Smartphones evolved into indispensable personal assistants, capable of managing both professional and personal aspects of users' lives.

As fourth-generation (4G) networks were rolled out globally, the speed and reliability of mobile internet increased exponentially. Video streaming, real-time social media interaction, and mobile gaming became ubiquitous. These advancements also paved the way for mobile payments and the rise of services like Uber, which leveraged GPS and mobile connectivity to create entirely new business models. The mobile phone had transcended its original purpose to become a versatile, multipurpose device ingrained in the fabric of modern life.

Currently, we are on the cusp of the next transformative phase with the deployment of fifth-generation (5G) technology. 5G promises ultra-high-speed internet, low latency, and the capacity to connect billions of devices, revolutionizing not just mobile telephony but various sectors. This advancement is expected to accelerate the growth of the Internet of Things (IoT), enabling smart cities, autonomous vehicles, and advanced healthcare solutions. Mobile phones equipped with 5G capabilities will not only enhance existing applications but will also facilitate new technologies we are only beginning to imagine.

Moreover, innovations in hardware continue to push boundaries. Foldable screens, advanced biometric security features, and integration with artificial intelligence have all become focal points for current and future mobile phone designs. These devices are now seen as extensions of our identities, capable of mirroring our preferences, routines, and even emotions through advanced algorithms and machine learning.

As we look ahead, the potential applications of future mobile phones seem limitless. Augmented reality (AR) and virtual reality (VR) are poised to become mainstream, creating immersive experiences that blend digital and physical worlds. Imagine using your mobile phone for virtual meetings where participants appear as holograms right in your living room, or navigating cities with AR overlays that provide real-time information about your surroundings. These scenarios are not far-fetched; they are the next steps in the ongoing evolution of mobile technology.

In tandem with technological advancements, mobile phones have also become more accessible to people across the globe. Emerging markets have seen rapid adoption of mobile technology, providing vital connectivity and access to information and services. Mobile banking, for example, has transformed financial inclusion by enabling millions of people without traditional banking access to manage their finances and conduct transactions securely.

The environmental impact and ethical considerations of mobile phone production and disposal have become increasingly important. The industry is taking steps to address issues like e-waste, resource consumption, and fair labor practices. Companies are exploring sustainable materials, improved recycling methods, and more ethical supply chains to mitigate the ecological footprint of mobile technology.

In conclusion, the evolution of mobile phones encapsulates a fascinating journey of technological innovation, societal transformation, and ever-expanding possibilities. From bulky devices with limited reach to sleek, multifaceted smartphones equipped with cutting-edge technology, mobile phones have radically altered how we communicate, work, and live. As new frontiers like 5G, AR, and AI emerge, the trajectory of mobile technology promises to continue reshaping our world in profound and unpredictable ways. The essence of this evolution lies in its relentless drive to connect us closer, faster, and more efficiently, illuminating the path toward an interconnected future brimming with potential.

Mobile Internet and Apps

The synergy between mobile internet and apps has fundamentally altered how we interact with the world, and it's a testament to human ingenuity. To appreciate the profound impact of these technologies, we need to explore the journey from their origins to their ubiquity in modern life. The evolution wasn't just a series of upgrades; it was a revolution.

The mobile internet began to take shape in the late 1990s and early 2000s. Initially, it was a slow and arduous experience. Early mobile networks like 2G focused solely on voice communication and text messages. When 3G networks were introduced, they marked the first significant leap towards mobile internet, offering data speeds that made web browsing, albeit sluggishly, possible. It was the dawn of a new era where connectivity wasn't confined to wires and desks anymore.

With the advent of 4G networks, mobile internet truly came of age. High-speed connectivity allowed for the seamless streaming of high-definition videos, instantaneous downloads, and the rise of complex mobile applications. This change wasn't merely technical; it was transformative. People could now access vast amounts of information and services in real-time, no matter where they were. The implications for global connectivity, business, education, and everyday life were enormous.

Parallelly, the development of smartphones was integral to this transformation. Devices like the iPhone and Android-based phones became powerful mini-computers, capable of running sophisticated applications. These apps leveraged the enhanced mobile internet capabilities to offer functionalities that were previously unimaginable on mobile devices.

One of the first app revolutions came in the form of social media. Platforms like Facebook, Twitter, and Instagram brought social connectivity to our fingertips. They transformed how we communicate, share, and consume information. The integration of mobile cameras with these apps enabled not just communication through text, but through images and videos, giving rise to a new era of digital storytelling.

But social media was just the beginning. The rise of mobile apps influenced every sector from banking to healthcare. Mobile banking apps allowed users to manage their finances with unprecedented ease and security. No longer did banking require a trip to a physical branch; a few taps on a mobile device sufficed.

Healthcare, too, witnessed a revolution. Mobile health apps, or mHealth apps, enabled users to monitor their well-being, access medical information, and even consult with doctors remotely. This democratization of healthcare knowledge and services was especially beneficial in remote and underserved areas, where traditional healthcare infrastructure was lacking.

Education was yet another field profoundly impacted by mobile apps. Platforms offering online courses, educational games, and interactive learning methods made education more accessible and engaging. Knowledge was no longer confined to traditional classrooms; it was available to anyone with a mobile device.

The proliferation of mobile apps also revolutionized commerce. E-commerce platforms like Amazon and eBay became more accessible, empowering consumers to shop anywhere, anytime. Mobile payment solutions, including Apple Pay and Google Wallet, further simplified transactions, making shopping more seamless than ever.

Entertainment found a new playground in mobile apps. Streaming services like Netflix, Spotify, and YouTube changed how we consume media. No longer were we tied to TV schedules or computer desks; our entertainment was mobile, ready to go wherever we went.

Business and productivity apps transformed work cultures across the globe. Apps for project management, communication, and productivity allowed for remote working and collaboration, heralding the era of the digital nomad. Tools like Slack, Trello, and Zoom became integral, especially highlighted during the COVID-19 pandemic, which underscored the necessity and utility of mobile technology in maintaining business continuity.

The gaming industry, too, experienced an upheaval. Mobile games became popular, transitioning from simple puzzles and arcade games to sophisticated, multi-player experiences. Games like Pokémon Go demonstrated the potential of augmented reality, engaging users in unprecedented ways that blended the digital and physical worlds.

Another critical area where mobile internet and apps have made substantial strides is transportation. Ride-hailing apps like Uber and Lyft revolutionized how we navigate urban spaces. These platforms offered convenience, often providing cheaper and quicker alternatives to traditional taxi services. Additionally, apps for public transportation schedules, bike rentals, and car-sharing aimed at reducing urban congestion and promoting sustainability.

Moreover, mobile apps have fostered community and social justice initiatives. Apps that facilitate citizen journalism, crowd-sourced data collection for scientific research, and platforms promoting civic engagement have empowered individuals to contribute to the greater good. Technology has acted as a great equalizer, giving voice to the voiceless and enabling grassroots movements to gain momentum rapidly.

As we delve deeper into the fabric of everyday life, it's evident that mobile internet and apps have woven themselves into our very existence. They've altered our routines and reshaped industries, demonstrating the endless potential of human creativity when harnessed through technology.

The future holds even more promise. As 5G technology becomes ubiquitous, the potential for more immersive and instantaneous app experiences becomes closer to reality. Augmented and virtual reality applications, enhanced by the low latency and high speeds of 5G, are set to revolutionize sectors such as healthcare, education, and entertainment even further. Smart cities, driven by interconnected devices and apps, promise to make our urban environments more efficient and livable.

It's not just about convenience and efficiency; it's about possibility and capability. Mobile internet and apps have made the world more accessible, breaking down barriers of distance and time. They've empowered individuals and communities, offering tools that were once the preserve of the privileged. The journey of mobile technology is a testament to humanity's incessant quest for progress, reflecting our collective ingenuity and drive towards continual improvement.

In conclusion, the impact of mobile internet and apps is profound and far-reaching. They symbolize a remarkable chapter in the story of human technological advancement, highlighting our ability to innovate and adapt. As we continue to navigate through this rapidly evolving landscape, it's clear that mobile technology will remain at the forefront of pushing boundaries and unlocking new frontiers in our interconnected world.

CHAPTER 18:
SOCIAL MEDIA

From its humble beginnings on message boards and early social networking sites, social media has transformed the way we communicate, share ideas, and engage with the world. The evolution from platforms like MySpace and Friendster to giants such as Facebook, Twitter, and Instagram signifies more than just a technological shift; it's a profound change in human interaction. Social media has bridged geographical gaps, creating virtual communities where ideas can proliferate and movements can gain momentum with unprecedented speed. Yet, it has also introduced complexities around issues like privacy, misinformation, and the psychology of online engagement. As a reflection of our society's drive for connection and self-expression, social media exemplifies the dual-edged nature of technological progress, presenting both incredible opportunities and poignant challenges. As we navigate this ever-evolving digital landscape, the implications for our social structures, personal identities, and global connectivity continue to unfold, shaping the very fabric of contemporary life.

From Message Boards to Facebook

Imagine a time when the Internet was a far cry from the densely interconnected web we navigate today. Back in the early days, social interactions online were confined to message boards and forums. These platforms were simple, often clunky, but they laid the groundwork for an unprecedented shift in how humans communicate.

Message boards, also known as forums, were some of the earliest forms of online social interaction. Literally serving as "boards" where users could post messages, these platforms allowed people from around the world to share information, discuss topics of interest, and form virtual communities. They were organized into threads, where posts on similar topics were grouped together. This seemingly rudimentary system fostered extended discussions that could span days, weeks, or even years.

In essence, message boards were the first large-scale digital public squares. Let's illustrate it with a bit more context. Imagine a group of gardening enthusiasts scattered across the globe. Before the Internet, these individuals would have had limited means of sharing their tips and tricks with each other.

With the advent of message boards, however, they suddenly had a virtual space to congregate. Threads would buzz with activity, filled with titles like "Best Organic Fertilizers" or "Tips for Growing Roses." Members would log in eagerly to see responses to their posts and engage in meaningful exchanges. The beauty of these platforms lay in their ability to bring together like-minded individuals, creating microcosmic societies where distance was irrelevant.

As time progressed, the landscape began to shift. Internet speeds increased, graphical interfaces improved, and a new type of platform started to emerge—social networking sites. Unlike message boards, where conversations were organized around topics, social networks focused on individuals. Users had profiles, could post status updates, and follow other users in a more fluid, less topic-constrained environment.

The watershed moment in this evolution came with the launch of Facebook in 2004. Initially created by Mark Zuckerberg as a college networking tool, Facebook rapidly expanded beyond university campuses. With its polished interface, real-time updates, and multifaceted interaction options, Facebook offered an entirely new level of user engagement. It wasn't just about connecting with friends; it was about sharing one's life in a dynamic, multimedia format.

Facebook did something revolutionary—it created a sense of immediacy and continuity. Real-time updates meant that you were aware of what was happening in your friends' lives almost as it unfolded. Photos, videos, and status updates combined to provide a multi-sensory experience, unlike the text-heavy message boards of old.

But the transformation wasn't merely cosmetic. Facebook functioned on a more sophisticated level, implementing algorithms to personalize user experience. These algorithms analyzed user behavior to prioritize content that was most likely to engage them, making the platform irresistibly addictive. This was the beginning of the social media phenomenon as we know it today: personalized, immediate, and immensely engaging.

While Facebook may have been the torchbearer, it wasn't alone for long. Other platforms like Myspace, Twitter, and LinkedIn emerged, each carving out their own niche. Myspace initially rivaled Facebook and became particularly popular for music sharing. Twitter introduced a novel concept of microblogging, allowing users to share succinct updates or "tweets" in real time. LinkedIn focused on professional networking, creating a valuable space for career-oriented connections.

These transformations had a profound impact on how society communicates, reshaping not just the medium but also the message. Social media democratized information dissemination, allowing anyone with an Internet connection to become a content creator, journalist, or influencer. News, opinions, and trends began to spread with unprecedented speed and reach, changing the landscape of journalism and public discourse.

The implications go even deeper. Online movements for social justice, political campaigns, and community organizing found fertile ground on social media platforms. Hashtags like #MeToo and #BlackLivesMatter became rallying points for change, facilitated by the global connectivity and immediacy that platforms like Facebook, Twitter, and Instagram provided.

However, the meteoric rise of social media also brought challenges. Issues such as privacy concerns, misinformation, and cyberbullying began to surface, highlighting the dark side of this newfound connectivity. Algorithms designed to maximize engagement sometimes had unintended consequences, such as creating echo chambers or spreading fake news.

Despite these challenges, it's undeniable that the journey from message boards to Facebook represents a pivotal chapter in the history of human communication. It's a journey that exemplifies how technology can reshape society, for better or for worse. The evolution of these platforms has mirrored our collective trajectory—from simple, interest-based communities to complex networks that integrate virtually every aspect of our lives.

Today, as we scroll through our feeds, post updates, and engage in online discourse, we're part of a continuum that began with those humble message boards. By understanding this evolution, we can better appreciate the immense potential—and responsibility—that comes with our digital interactions.

Moving forward, it will be fascinating to see how social media will continue to evolve. Will it become more immersive, integrating elements of virtual and augmented reality? Will new platforms emerge that prioritize user well-being and privacy more robustly? Only time will tell. One thing is certain, though: the innovation sparked by those early message boards will continue to reverberate, influencing not just how we communicate, but how we connect, organize, and even think.

Impact on Communication and Society

Social media has fundamentally altered the landscape of human communication in ways that were unimaginable even a few decades ago. With just a few keystrokes or taps on a screen, individuals can now share their thoughts, experiences, and ideas with a global audience. This unprecedented ability to connect with others has brought about both profound benefits and significant challenges.

One of the most significant impacts of social media on communication is its ability to transcend geographical boundaries. Friends and family members separated by oceans or continents can maintain close relationships through platforms like Facebook, Instagram, and WhatsApp. In an era where migration and global mobility are common, these tools provide an invaluable means of maintaining personal connections. However, while social media can enhance these relationships, it can also create a sense of false intimacy, as online interactions often lack the depth and nuance of face-to-face conversations.

The democratization of information sharing is another major impact of social media. Traditional gatekeepers of information, such as news organizations and academic institutions, no longer hold a monopoly on knowledge dissemination. Anyone with an internet connection can now contribute to the global dialogue on any subject. This has led to a more inclusive and diverse discourse, where previously marginalized voices can be heard. Nonetheless, this democratization comes with the downside of spreading misinformation, as not all shared content is subjected to rigorous fact-checking processes.

In the realm of social activism, social media has proven to be a double-edged sword. On one hand, it has empowered movements like #MeToo, Black Lives Matter, and climate change activism, allowing organizers to mobilize supporters quickly and spread their message widely. On the other hand, the same platforms also facilitate the rapid dissemination of disinformation and hate speech. The viral nature of social media means that falsehoods or inflammatory statements can spread far more quickly than factual information, complicating efforts to maintain a well-informed public.

From a psychological perspective, the impact of social media on individuals is a topic of ongoing debate. Some studies suggest that the constant exposure to curated lives and idealized images can lead to feelings of inadequacy, anxiety, and depression. The phenomenon of "FOMO" (Fear of Missing Out) puts additional pressure on individuals to continuously engage with social media to stay updated, often at the expense of their mental well-being. Conversely, for many people, social media provides a crucial sense of community and belonging, especially for those who feel isolated in their offline lives.

Moreover, social media has revolutionized the way businesses communicate with consumers. Companies now use these platforms for marketing, customer service, and brand building. The ability to engage directly with their audience in real-time has transformed traditional marketing strategies. Brands that successfully leverage these tools can build loyal, engaged customer bases more effectively than through traditional advertising methods. However, this direct line of communication also means that brands are more vulnerable to public relations crises, as negative feedback can spread rapidly and affect reputations almost instantaneously.

Educational institutions and educators have also harnessed the power of social media for teaching and learning. Online discussions, live streaming of lectures, and collaborative projects made possible through these platforms extend learning beyond the classroom walls. Social media fosters peer-to-peer learning, where students can share resources and insights readily. Yet, this integration is not without challenges, including issues of digital distraction and unequal access to technology, which can exacerbate educational inequalities.

Social media's influence on politics and democracy is perhaps one of its most contentious impacts. The platforms have become arenas for political discourse, campaigning, and even electoral interference. Political candidates and parties use social media extensively to reach voters, share policy positions, and shape public opinion. While this can enhance democratic engagement by making information more accessible, it also opens the door to manipulation tactics like echo chambers and filter bubbles, where individuals are exposed primarily to viewpoints that reinforce their own beliefs, thereby polarizing society.

On the international stage, social media plays a role in diplomacy and global relations. Political leaders and governments use platforms like Twitter and Facebook for official communications, to manage crises, and to reach international audiences. This instant, unfiltered communication can bypass traditional media outlets and offer a direct channel to the public. However, it also introduces risks, as the informal nature of social media can lead to misunderstandings or diplomatic gaffes with far-reaching consequences.

The realm of cultural exchange has also been transformed by social media. Distinct cultural practices, art forms, and traditions are shared across borders with ease, fostering greater global understanding and appreciation. Yet, this exchange is not always equal, and there are concerns about cultural appropriation and the erosion of local cultures under the influence of dominant ones. The free flow of information can lead to a homogenization of cultures, where unique local traditions risk being overshadowed by global trends.

Social media has even impacted the way we perceive and consume entertainment. From viral videos to online gaming communities, these platforms have democratized content creation and consumption. Anyone can become a content creator, gaining massive followings and even turning hobbies into lucrative careers. The entertainment industry has had to adapt to these changes, finding new ways to engage audiences who have shorter attention spans and more options than ever before. This shift has also led to issues like digital piracy and questions about the value of creative work in an age where content is abundant and often free.

The ethical challenges presented by social media are manifold. Issues of privacy, data security, and algorithmic transparency are at the forefront of ongoing debates. Users often share personal information without fully understanding the extent to which it can be exploited by corporations or even malicious actors. Data breaches and the misuse of personal information have led to calls for stricter regulations and more robust protections for users. Furthermore, the algorithms that curate content on these platforms raise questions about bias and manipulation, as they can influence public perception and behavior in subtle but powerful ways.

In conclusion, social media's impact on communication and society is vast and multifaceted. It has opened up incredible opportunities for connection, information dissemination, and collective action. At the same time, it presents significant challenges that require careful consideration and ongoing vigilance. As we continue to navigate this digital landscape, the lessons we learn will undoubtedly shape the future trajectory of our communication practices and the very fabric of our society.

CHAPTER 19:
RENEWABLE ENERGY

As humanity faces ever-growing environmental challenges, the significance of renewable energy solutions becomes startlingly clear. Solar and wind power, among the most promising prospects, harness natural forces to generate electricity without depleting resources or emitting greenhouse gases. These technologies have evolved from niche applications to mainstream adoption, driven by advances in efficiency and cost reductions. Yet, it's not just about energy production; innovations in sustainability, such as improved energy storage systems and smart grids, promise to revolutionize the way we consume and conserve power. Our transition to renewables thus signifies more than a shift in energy sources—it's a profound step towards a sustainable future, embracing the ingenuity and resilience that have always characterized human progress.

Solar and Wind Power

As we journey through the saga of human progress, renewable energy sources like solar and wind power stand as towering examples of our collective ingenuity and forward-thinking attitude. From harnessing the might of the sun to capturing the kinetic energy of wind, these technologies aren't just scientific marvels but also hope-filled answers to pressing global challenges.

Solar energy, in particular, has tapped into the most ancient of power sources — the sun. Through the ingenious use of photovoltaic

(PV) cells, we convert sunlight directly into electricity. These cells, often made from silicon, create an electric current when exposed to sunlight. This process, straightforward as it may seem, is a symbol of human tenacity and scientific excellence. Early developments in PV technology faced numerous challenges, including inefficiency and high production costs. However, relentless research and innovation have driven costs down drastically while increasing efficiency. From large-scale solar farms that fuel entire communities to small rooftop installations powering individual homes, solar power has become an integral part of the energy mosaic.

Wind power, on the other hand, harnesses a different element of nature. Wind turbines convert the kinetic energy from wind into mechanical power, which can then be transformed into electricity via generators. The concept is simple, yet the execution is nothing short of grand. Modern wind turbines are feats of engineering, with some standing taller than the Statue of Liberty. These gigantic structures, often grouped in wind farms, are capable of providing enormous amounts of energy without emitting greenhouse gases. Their potential is most evident in regions with strong and consistent wind patterns, but technological advancements are increasing their viability in more diverse locales.

Both solar and wind power have the remarkable ability to decentralize the energy grid. Unlike traditional power plants, which often require extensive infrastructure for distribution, renewable energy installations can be set up at various scales and locations. This not only increases energy accessibility but also enhances resilience against natural disasters and other disruptions. Moreover, these renewable sources create opportunities for communities to become energy independent, reducing reliance on fossil fuel imports and promoting local economic growth.

The environmental benefits are equally profound. Solar and wind power produce negligible carbon emissions once operational. Unlike coal or natural gas plants, renewables don't pollute our air or water. This contributes to mitigating climate change and improves public health by lowering the prevalence of respiratory issues and other pollution-related health problems. By transitioning to renewables, societies can achieve cleaner air, healthier populations, and a more stable climate.

Economically, the renewable energy sector is a burgeoning frontier. Jobs in solar and wind have been growing exponentially, outpacing those in many traditional energy sectors. From manufacturing PV cells to maintaining wind turbines, this growing industry offers a plethora of job opportunities. Additionally, as technology improves, the cost of renewable energy continues to decrease, making it not just an environmentally friendly choice but also an economically viable one.

Innovation in storage technology further amplifies the potential of solar and wind power. Energy storage systems like lithium-ion batteries allow for the capture and use of energy even when the sun isn't shining or the wind isn't blowing. These advancements are crucial for overcoming one of the major limitations of renewable energy: intermittency. With effective storage solutions, renewable energy can provide a consistent and reliable power supply, further integrating into the existing energy grid.

Government policies and incentives also play a pivotal role in the adoption and proliferation of solar and wind power. Subsidies, tax incentives, and feed-in tariffs have driven significant investment in these technologies. Countries around the world are setting ambitious renewable energy targets to reduce carbon footprints and transition to greener economies. These policies not only encourage the adoption of existing technologies but also spur further innovation and development in the renewable energy sector.

The journey of solar and wind power is also deeply intertwined with the leveraging of big data and IoT (Internet of Things). Smart grids, powered by data analytics and real-time monitoring, optimize the distribution and consumption of renewable energy. Predictive maintenance of wind turbines and solar panels using IoT-enabled sensors helps reduce downtimes and improve efficiency. The integration of these cutting-edge technologies signifies a future where renewable energy systems are smarter, more interactive, and incredibly efficient.

Despite these advancements, challenges remain. Initial installation costs can be high, especially in areas lacking the necessary infrastructure. Furthermore, the aesthetic impact of large installations and their effects on local wildlife, such as birds and bats around wind farms, are concerns that require careful planning and innovation to mitigate. Nevertheless, the momentum behind renewable energy, bolstered by immense technological progress and supportive policies, suggests that these issues are surmountable.

Looking ahead, the transformative potential of solar and wind power is undeniable. As we stand on the cusp of a new era in energy production, these renewable sources offer a glimpse into a future where clean, sustainable, and abundant energy is within reach. Our journey from early tool-making to advanced technologies has always been fueled by a quest to understand and harness the forces of nature. Solar and wind power underscore this narrative, representing not just technological prowess but also a commitment to a more sustainable and equitable world.

Indeed, the proliferation of solar and wind power technologies embodies a harmonious melding of human ingenuity, environmental stewardship, and economic opportunity. As we continue to innovate and improve upon these technologies, we come ever closer to realizing a world powered by clean and renewable energy. This isn't merely a hoped-for future; it's a dynamic and rapidly advancing reality that could redefine how we perceive energy consumption and its impact on our planet.

Innovations in Sustainability

The quest for renewable energy is not just an exploration of new technologies but also an endeavor rooted in sustainability. In the ever-increasing demand for energy, innovations in sustainability are steering the direction towards a cleaner, greener future. These innovations are ensuring that we can harness the power of the Earth without depleting its resources, making it possible to meet our current needs while preserving the ability of future generations to meet theirs.

One of the paramount innovations in sustainability is the development of advanced solar panel technologies. Traditional photovoltaic cells have been around for decades, but recent advancements have significantly improved their efficiency and cost-effectiveness. Researchers are focused on multi-junction and bifacial solar cells that can capture sunlight from various angles and spectrums, resulting in higher energy output. Moreover, thin-film solar cells, made from materials like cadmium telluride and perovskite, are leading the charge due to their lower production costs and flexibility, enabling their integration into various surfaces and environments.

Similarly, the realm of wind energy has seen transformative innovations. Modern wind turbines are now towering structures that can harness more wind energy than their predecessors. Innovations such as bladeless turbines, which use oscillation to generate power, promise to mitigate some of the environmental concerns related to traditional wind farms, like harm to bird populations and noise pollution. Enhanced materials and aerodynamic designs for turbine blades have also contributed to increased efficiency and reduced maintenance costs, making wind farms a more viable component of our energy grids.

The development of energy storage systems has been a critical innovation for the feasibility of renewable energy sources. Intermittency is a significant challenge with solar and wind power, as they depend on weather conditions and time of day. Advanced battery technologies, such as lithium-ion and emerging solid-state batteries, are making it possible to store surplus energy generated during peak production times and release it when needed. Innovations in battery management systems ensure optimal performance and longevity, providing a reliable backup for renewable energy sources.

In addition to energy production and storage, grid management systems have evolved to accommodate the influx of renewable energy. Smart grids, using advanced computer algorithms and real-time data analysis, can autonomously manage the distribution of power. These grids can predict energy demand, detect faults, and reroute electricity accordingly, enhancing efficiency and minimizing disruptions. Moreover, microgrid technology supports localized energy production and usage, allowing communities to be more self-sufficient and resilient to broader grid failures.

The integration of smart technologies is playing a critical role in sustainability. The Internet of Things (IoT) devices, encompassing smart meters, smart appliances, and adaptive lighting, enable users to monitor and reduce their energy consumption. Real-time data collection and analysis facilitate more informed decisions, from household levels to industrial scales. These technologies empower consumers to actively participate in energy conservation efforts, promoting a collective approach towards a sustainable future.

In urban settings, sustainable innovations are transforming the concept of cities. Green buildings, designed with energy efficiency at the forefront, utilize materials and construction techniques that reduce environmental impact. Features such as green roofs, which provide insulation and reduce heat islands, and energy-efficient HVAC systems, are becoming standard practice. Additionally, the application of solar panels, wind turbines, and rainwater harvesting systems in buildings creates a microcosm of renewable energy right in the heart of cities.

Another remarkable innovation lies in the realm of bioenergy. Bioenergy involves converting organic materials, such as agricultural waste, algae, and even municipal waste, into usable energy. Technologies like anaerobic digesters and advanced biofuel processing techniques are maximizing the energy output from these materials. Bioenergy not only provides a renewable source of power but also offers a solution to waste management, turning potential pollutants into valuable resources.

The agricultural sector, closely tied to bioenergy, is also benefitting from innovations in sustainability. Practices like precision farming use GPS and IoT technology to optimize the use of resources like water, fertilizer, and pesticides. By monitoring soil conditions and plant health in real-time, farmers can make data-driven decisions that reduce waste and increase yield. This approach ensures that agriculture can meet the increasing global food demand without compromising environmental integrity.

Even ocean-based renewable energy is making waves in the field of sustainability. Technologies like tidal and wave energy convert the kinetic motion of ocean currents and waves into electricity. These systems, still in their nascent stages, hold immense potential due to the predictability and consistency of ocean movements. Innovations in underwater turbines and wave energy converters are steadily improving efficiency, durability, and the environmental compatibility of this form of renewable energy.

Public policy and international collaboration are critical to advancing these innovations in sustainability. Governments worldwide are implementing incentives, subsidies, and regulations that foster renewable energy development. The Paris Agreement, for example, highlights global efforts to coordinate climate action, setting targets for reducing carbon emissions and promoting green technologies. Collaborative initiatives, such as international research grants and technology-sharing agreements, expedite the development and deployment of sustainable energy solutions.

On the consumer side, sustainable lifestyle choices are gaining traction. From adopting solar panels on rooftops to driving electric vehicles, individuals are making significant inroads into reducing their carbon footprint. Community-shared energy projects, where neighborhoods collectively invest in renewable energy systems, are also becoming popular. Such grassroots movements are pivotal, as they demonstrate the collective impact of individual actions on achieving broader sustainability goals.

Lastly, education and awareness are crucial components of the sustainability paradigm. Programs that educate the public and future generations about the benefits and practicalities of renewable energy foster a culture of environmental stewardship. Schools, universities, and community organizations play pivotal roles in disseminating knowledge and encouraging sustainable practices. By raising awareness and building competence in green technologies, society becomes better equipped to tackle the challenges of energy sustainability.

In conclusion, innovations in sustainability are multifaceted and dynamic. They encompass technological advancements, improved practices, and systemic changes that collectively lead to a more sustainable future. Renewable energy, as part of this broader effort, is pivotal in reducing our carbon footprint and ensuring the well-being of the planet. Through continued innovation and collaborative efforts, we can harness the Earth's natural resources responsibly, creating a legacy of sustainability that future generations can build upon.

CHAPTER 20:
BIOTECHNOLOGY

Biotechnology stands at the crossroads of biology and technology, ushering in a new era of transformative possibilities. With the advent of genetic engineering and tools like CRISPR, we've cracked open the very code of life itself. These innovations aren't just academic; they translate directly into tangible benefits in fields like medicine and agriculture. Imagine crops engineered to withstand drought or diseases cured at their genetic roots. The implications are profound. We are not just observers but active participants in sculpting the future, leveraging the power of biology to solve some of humanity's most pressing problems. As we journey through this chapter, we will explore the awe-inspiring advancements that biotechnology has brought to our world, forever changing the landscape of what is possible in science and medicine.

Genetic Engineering and CRISPR

Biotechnology has profoundly transformed human capability, but few elements have captured public imagination and scientific potential as vividly as genetic engineering and CRISPR. These advancements are not merely technological innovations; they represent a fundamental shift in how we can understand, manipulate, and harness the very building blocks of life itself.

Genetic engineering began in the latter half of the 20th century with the development of recombinant DNA technology. Simply put, scientists learned how to cut and paste genes from one organism to another, leading to the creation of genetically modified organisms (GMOs). By introducing foreign genes, scientists could endow organisms with new traits, whether it was pest resistance in crops or the production of insulin in bacteria. The implications were revolutionary: improved agricultural yields, novel pharmaceuticals, and even the potential to eradicate hereditary diseases.

Yet, as transformative as recombinant DNA technology was, it also had its limitations. Traditional genetic engineering techniques were labor-intensive, costly, and often imprecise. Enter CRISPR—the tool that has redefined the possibilities of genetic engineering. Discovered in the early 2010s, CRISPR stands for "Clustered Regularly Interspaced Short Palindromic Repeats." This mouthful describes a powerful, precise, and relatively simple method for editing the genome. CRISPR uses an enzyme called Cas9, guided by a specifically designed RNA sequence, to make targeted cuts in DNA strands. The result? Scientists can easily add, remove, or alter genetic material with unprecedented precision.

What makes CRISPR revolutionary is its versatility and ease of use compared to earlier genetic engineering techniques. Labs around the world adopted CRISPR almost overnight due to its ability to perform complex gene edits with relative simplicity. One of the iconic early applications of CRISPR was its use in modifying the genes of human embryos, moving scientists closer to the once-science-fiction notion of 'designer babies'. However, the ethical considerations surrounding such applications are nothing short of labyrinthine.

The potential of CRISPR goes beyond human embryonic changes. Agricultural sectors have greatly benefitted as well. Scientists have used CRISPR to produce crops with enhanced nutritional value, better yield, and greater resistance to diseases and environmental stresses. Imagine a world where food security is no longer a concern, where staple crops can flourish even in arid climates or under attack by pests thanks to CRISPR-induced genetic fortification.

In medicine, the implications are equally astounding. Researchers are now exploring CRISPR as a means to directly treat genetic disorders. Conditions like cystic fibrosis, muscular dystrophy, and even some forms of cancer are susceptible to CRISPR-based treatments which aim to correct the underlying genetic mutations. These applications are still largely in the research phase, but clinical trials are underway, offering a glimpse into a future where genetic diseases could be eradicated before they manifest.

Moreover, CRISPR has been a game-changer in the field of microbial resistance. The technology enables the development of new antibiotics and the modification of microbial genes to combat antibiotic-resistant bacteria. Scientists have proposed the use of CRISPR to create gene drives—mechanisms that can spread particular genetic traits through populations at an accelerated rate. Such technology holds promise in controlling or even eliminating vectors for diseases like malaria by modifying the genes of mosquitoes.

Of course, the rapid advancement of CRISPR has prompted a robust dialogue around bioethics. The ability to edit genomes raises profound questions. How should we regulate such powerful technology? What are the long-term ecological and societal impacts? Who gets to decide what constitutes a 'desirable' genetic trait? These are not questions with easy answers, and they necessitate careful consideration from scientists, ethicists, policymakers, and the public alike.

One of the most compelling ethical considerations is the potential for CRISPR to exacerbate social inequalities. As with many groundbreaking technologies, there is a risk that access will be limited to those who can afford it, deepening the divide between rich and poor. Imagine a world where genetic enhancements are available only to a privileged few, leading to a new form of genetic aristocracy. Such scenarios aren't merely the stuff of dystopian fiction; they are real possibilities that must be addressed as CRISPR technology continues to evolve.

The future of genetic engineering and CRISPR is undoubtedly bright, but it is also fraught with challenges. As we stand on the cusp of a new era where the genome can be edited with ease, the ethical, social, and environmental implications demand our utmost attention. The power to rewrite the code of life holds immense promise but also mandates caution and thoughtful governance.

In summary, genetic engineering and CRISPR are at the forefront of biotechnology's transformative impact on humanity. They offer unprecedented capabilities, from disease eradication to agricultural innovation and beyond. Yet, they also require us to grapple with profound ethical dilemmas and a responsibility to ensure equitable and mindful application. As we venture deeper into this era of genetic mastery, the lessons we've learned from centuries of technological advancement must guide us: progress must be tempered with prudence, and innovation balanced by introspection.

Medicine and Agricultural Benefits

Biotechnology has bridged an unprecedented gap between our understanding of life itself and the ways we can manipulate it to better human health and agricultural productivity. This convergence of biology, technology, and innovation presents opportunities that were once confined to the realm of science fiction. Now, these advancements promise a future where genetic engineering and other biotechnologies can alleviate suffering, enhance the quality of life, and address pressing global challenges.

In medicine, the impact of biotechnology is nothing short of revolutionary. Consider the field of genetic engineering, which allows us to modify DNA to correct genetic disorders and even prevent certain diseases. CRISPR-Cas9, a groundbreaking gene-editing technology, has changed the landscape of biomedical research. It enables scientists to precisely target and edit specific genes, overcoming the limitations of previous methods. This holds enormous potential for treating conditions like cystic fibrosis, sickle cell anemia, and even certain cancers.

Beyond genetic engineering, biotechnology has also led to the development of personalized medicine. This approach tailors medical treatment to the individual characteristics of each patient. By analyzing genetic information, healthcare providers can prescribe treatments that are more effective and have fewer side effects. This shift from a one-size-fits-all model to a more customized approach is transforming patient care, making it more efficient and effective.

Moreover, biotechnology has significantly improved vaccine production. Traditional vaccines often took years to develop, but advances in biotech have streamlined this process. For instance, the rapid development of COVID-19 vaccines was made possible by mRNA technology, a triumph of biotechnological innovation. This is poised to change the way we respond to emerging infectious diseases in the future.

Biotechnology does not stop at medicine; its benefits spill over into agriculture as well. One of the most prominent applications is the development of genetically modified organisms (GMOs). These crops are engineered to be more resistant to pests, diseases, and harsh environmental conditions, thereby increasing agricultural productivity and food security. For instance, Bt cotton and Bt corn have been genetically altered to produce a toxin that repels or kills specific insects, reducing the need for chemical pesticides.

Apart from pest resistance, biotechnology also enables crops to tolerate extreme weather conditions, such as drought or saline soil. This is crucial as climate change poses a significant threat to food production worldwide. Drought-tolerant crops, for example, can maintain yields even during prolonged periods of low rainfall, ensuring that farmers' livelihoods and global food supplies remain stable.

In addition to GMOs, biotechnology has facilitated the development of biofortified crops. These are crops that have been enriched with essential vitamins and minerals to address nutrient deficiencies in populations that rely heavily on staple crops. An example is Golden Rice, which has been engineered to produce beta-carotene, a precursor of vitamin A. This innovation aims to combat vitamin A deficiency, which is prevalent in many developing countries and can lead to severe health problems, including blindness.

Another significant contribution of biotechnology in agriculture is the advancement of precision farming. This involves using biotechnological tools to optimize field-level management with regard to crop farming. Techniques like genetic mapping, molecular markers, and DNA fingerprinting allow for the identification of desirable traits in crops, leading to the creation of high-yield and high-quality varieties.

Furthermore, biotechnology enhances sustainable agricultural practices. Through the development of microorganisms and biofertilizers, soil health and fertility can be significantly improved. These biofertilizers provide plants with essential nutrients more efficiently and sustainably than traditional chemical fertilizers, promoting healthier growth and reducing the environmental impact of farming practices.

The benefits of biotechnology extend to animal husbandry as well. Genetic engineering and cloning technologies have enabled the production of healthier, more productive livestock. Animals can be genetically engineered to be resistant to certain diseases, reducing the need for antibiotics and minimizing the risk of zoonotic diseases that can jump from animals to humans. Cloning, although controversial, offers the potential to preserve endangered species and improve genetic lines in livestock.

Biotechnology also plays a role in the development of alternative protein sources, which are crucial for meeting the growing global demand for meat without further straining the environment. Innovations like lab-grown meat and plant-based meat substitutes are promising solutions that mimic the taste and texture of traditional meat while reducing the environmental footprint of meat production. These technologies could significantly lower greenhouse gas emissions, reduce land and water use, and decrease reliance on antibiotics.

In addition to these tangible benefits, biotechnology fosters collaboration and innovation across disciplines. It brings together biologists, chemists, engineers, and computer scientists to solve complex problems, driving interdisciplinary research and development. This collaborative approach is essential for tackling the multifaceted challenges we face in medicine and agriculture.

The potential of biotechnology is vast, but it also raises ethical and regulatory questions. Genetic modifications, whether in humans, animals, or plants, come with risks and uncertainties. It is crucial to have robust regulatory frameworks to ensure that biotechnological advancements are safe, ethical, and accessible to all. There is a need for ongoing dialogue among scientists, policymakers, and the public to navigate these challenges responsibly.

As we look to the future, the role of biotechnology in transforming medicine and agriculture is likely to grow even more profound. The ongoing research and development in this field promise solutions to some of the most pressing global issues, from preventing and curing diseases to ensuring food security in the face of climate change. These innovations hold the potential to not only improve quality of life but also to pave the way for a more sustainable and equitable world.

In conclusion, biotechnology stands at the forefront of human progress, offering unparalleled benefits in both medicine and agriculture. From the precise editing of genetic material to the development of resilient and nutritious crops, these advancements hold the promise of addressing some of the most pressing challenges of our time. Through careful stewardship and ethical considerations, we can harness the full potential of biotechnology to create a healthier, more sustainable future for all. This intersection of biology and technology exemplifies the boundless possibilities that arise when human ingenuity meets cutting-edge science.

CHAPTER 21:
SPACE EXPLORATION

Space exploration stands as one of humanity's most formidable and inspiring adventures, pushing the boundaries of what we know and what we dare to dream. From the exhilarating days of the Space Race, which pitted nations in a quest to reach the cosmos, to today's collaborative international missions, the quest to explore the universe has ignited our imagination and spurred unprecedented technological advancements. This era not only marks humanity's steps on foreign celestial bodies but also signifies the burgeoning potential of commercial ventures and private enterprises boldly entering the final frontier. With each satellite launch, rover landing, and space station experiment, we inch closer to unlocking the mysteries of our galaxy and perhaps, one day, establishing a presence beyond Earth. The journey of space exploration exemplifies our innate drive to explore, discover, and transcend our terrestrial limitations, paving the way for a future where the stars are not just points of light, but destinations.

The Space Race

The Space Race stands as one of humanity's most audacious competitions, driving technological innovation at breakneck speed. This era, primarily pitting the United States against the Soviet Union, wasn't just about achieving milestones in space; it was a reflection of geopolitical tensions, a display of technological prowess, and a testament to human curiosity and ambition.

169

In the aftermath of World War II, the two emerging superpowers, the United States and the Soviet Union, found themselves in a fierce rivalry. Both nations sought to prove their superiority not only on Earth but in the infinite expanse of space. Initially, the Space Race can be traced back to a specific starting point: the Soviet Union's successful launch of Sputnik 1 on October 4, 1957. This was the first artificial satellite to orbit the Earth, and it sent shockwaves around the globe. The unmistakable beep-beep of Sputnik reverberated in the hearts of millions, symbolizing both a technological triumph and a pressing need for the United States to catch up.

America's response was swift but fraught with setbacks. The creation of NASA in 1958 marked a significant step towards organizing the nation's efforts in space exploration. The early days were a mix of triumphs and embarrassing failures, notably the Vanguard TV3 explosion later nicknamed "Flopnik." However, this early struggle only galvanized American determination. Under the directive of President John F. Kennedy, the U.S. made a bold proclamation: to land a man on the moon and return him safely to Earth before the decade's end.

The Apollo program emerged from this ambitious directive. While the early Mercury and Gemini missions laid critical groundwork, such as performing the first American spacewalk and testing space rendezvous techniques, it was Apollo that elucidated NASA's ultimate goal. Each mission was a meticulously planned step, each accomplishment building towards the seminal moment of Apollo 11. When Neil Armstrong set foot on the lunar surface on July 20, 1969, with the words, "That's one small step for man, one giant leap for mankind," it was a collective exhale and a resounding victory for the United States. It symbolized not merely a win in the Space Race but a win for all humanity.

However, it's crucial to acknowledge the significant breakthroughs on the Soviet side. The USSR wasn't merely the initiator with Sputnik;

they achieved many other firsts. Yuri Gagarin's pioneering flight on April 12, 1961, made him the first human to journey into outer space and safely return, inspiring awe worldwide. The Soviets also achieved the first spacewalk by Alexei Leonov and paved the way for prolonged human presence in space with the launch of the first space station, Salyut 1, in 1971.

The Space Race wasn't solely about achieving singular milestones; it catalyzed tremendous advancements in various technologies. Miniaturization, the development of computer systems, advancements in telecommunications, and novel engineering techniques sprouted from the needs of space exploration. In essence, the intense competition on the cosmic frontier accelerated technological innovation at a rate that would have been unimaginable in a different context.

As we examine the Space Race, it is important to view it not simply as a contest won by America but as a cooperative achievement in human history. After the Apollo triumph, the spirit of collaboration began to replace the spirit of competition. The Apollo-Soyuz Test Project in 1975, where an American Apollo spacecraft docked with a Soviet Soyuz capsule, symbolized a groundbreaking thaw in Cold War hostilities and a commitment to shared progress.

Beyond the immediate technological and scientific benefits, the Space Race had profound psychological and cultural impacts. For millions of people around the world, the images of Earth from space—a fragile "blue marble" floating in the void—altered perspectives on our planet, environment, and the interconnectedness of all human endeavors. This was the genesis of the environmental movement and a broader understanding of global interdependence.

Educational sectors also reaped long-term benefits. With the spotlight on science, technology, engineering, and mathematics (STEM), nations recognized the need to foster the next generation of innovators. Governments poured resources into educational programs, inspiring countless young minds to pursue careers in science and technology. The ripples of this investment are still felt today, as we continue to push the boundaries of what's possible.

Moreover, the Space Race demonstrated that with ambition, collaboration, and innovation, humanity could achieve the seemingly impossible. It became a metaphor for pushing beyond our collective limits. In this race, rockets were not just vessels of national pride but symbols of human aspiration, ingenuity, and our relentless drive to explore the unknown.

Today, we build on the foundation laid during the Space Race. The International Space Station (ISS), a colossal cooperative effort involving over fifteen nations, stands as a testament to how far we've come from those early days of intense rivalry. Institutions and companies around the globe collaborate on projects aiming for Mars and beyond. Space tourism, once the realm of science fiction, is on the cusp of becoming a reality, with private enterprises like SpaceX and Blue Origin leading the way.

In reflecting on the Space Race, we are reminded of the myriad ways technological competition can spur human advancement. We look back not just at a tale of two superpowers, locked in a duel for dominance, but at a chapter in human history that showed what is possible when creativity and ambition meet technological potential.

As we chart the future course of space exploration, it's essential to take lessons from the Space Race. Collaborative frameworks, sustainable innovations, and inclusive advancements can drive the next era of discovery. Emerging technologies, such as artificial intelligence, robotics, and biotechnology, promise to reshape our understanding of space and its utilization, reminiscent of the transformative impacts experienced during the Space Race.

While we've come a long way since the first beeps of Sputnik echoed through space, the journey is far from over. Looking ahead, our strides into the cosmos will continue to reflect our quest for knowledge, our desire to overcome challenges, and our boundless enthusiasm to see what lies beyond the horizon.

Commercial and International Ventures

The era of space exploration has extended far beyond the boundaries of governmental programs. As humanity's gaze turned towards the stars, so too did the interests of private enterprises and international coalitions. This section delves into the burgeoning commercial sector of space travel and the intricate tapestry of international collaborations that have formed an essential part of our quest to explore the cosmos.

The advent of commercial ventures in space exploration represents a paradigm shift reminiscent of the age of exploration on Earth. Private companies, once relegated to the role of suppliers for national space agencies, have positioned themselves at the forefront of launching and sustaining missions. Visionaries like Elon Musk with SpaceX and Jeff Bezos with Blue Origin have demonstrated that the private sector can indeed carve out a significant niche in space. SpaceX's reusable rockets have not only reduced the cost of accessing space but have also ignited a competitive spirit, pushing boundaries previously thought to be insurmountable.

Similarly, Blue Origin's mission to build a road to space, with aspirations of enabling humanity to live and work in space, illustrates how private industry is dreaming bigger. Their New Shepard suborbital vehicle has made numerous successful flights, signaling the dawn of space tourism. These ventures have opened up unprecedented opportunities for non-governmental entities to conduct research, develop technologies, and even offer commercial flights to the public.

Space tourism, in particular, has transitioned from the realm of science fiction to practical reality. Companies are working tirelessly to create experiences where ordinary individuals can travel to the edge of space, experiencing weightlessness and marveling at the curvature of the Earth. Virgin Galactic, founded by Richard Branson, is another critical player in this nascent industry. Their vision to democratize access to space has led to the development of SpaceShipTwo, an air-launched spacecraft designed for suborbital spaceflight.

But the commercial sector's involvement doesn't stop there. Industrial applications in space are becoming increasingly viable. For instance, companies like Planet Labs and Maxar Technologies are revolutionizing remote sensing and geospatial analytics through fleets of small satellites. These satellite constellations provide critical data for environmental monitoring, urban planning, and disaster response.

Moreover, the International Space Station (ISS) serves as a cornerstone of commercial and international collaboration. Since its inception, the ISS has been a testament to what can be achieved when nations put aside their differences for the sake of scientific advancement. It's a unique laboratory where commercial entities can conduct experiments in microgravity, contributing valuable insights that could have profound implications on Earth. NASA's Commercial Crew Program, which partners with private companies to deliver astronauts to the ISS, exemplifies this collaboration.

In addition to fostering commercial ventures, space agencies worldwide have cultivated significant international partnerships. The European Space Agency (ESA), Roscosmos, NASA, the Canadian Space Agency (CSA), and JAXA (Japan Aerospace Exploration Agency) have all played integral roles in various missions, often collaborating to optimize resources and share expertise. The successful deployment of the James Webb Space Telescope is an excellent example of such cooperation, showcasing how combining talents and resources from multiple countries can lead to groundbreaking achievements.

As we venture farther into space, such cooperation becomes ever more critical. The ambitious Artemis program, spearheaded by NASA, aims to land the next humans on the Moon and establish a sustainable presence there by the end of this decade. This program not only relies on private sector involvement but also heavily on international partners. For instance, ESA is contributing the European Service Module for the Orion spacecraft, and agencies from multiple countries are collaborating on the Gateway, a space station intended to orbit the Moon, which will serve as a staging point for deep-space missions, including crewed missions to Mars.

The political and economic implications of these collaborations cannot be understated. They foster a sense of unity and shared purpose among participating nations, reminding us that space is a global commons. However, they also require navigating a complex web of international laws and agreements, like the Outer Space Treaty of 1967, which serves as the foundation of international space law, establishing that space exploration must benefit all humankind.

Another intriguing offshoot of international ventures in space is the race to harness extraterrestrial resources. The potential for mining asteroids, extracting water from lunar ice, or even capturing solar energy from space and transmitting it to Earth are subjects of serious research and development. NASA's Artemis Accords, an international agreement aimed at establishing a framework for cooperative lunar exploration and beyond, underscores the importance of these activities. Companies like Planetary Resources and Deep Space Industries have laid the groundwork for what could become a thriving space resource extraction industry.

These ambitious goals have spurred innovations that trickle down to benefit our daily lives. For example, advances in robotics and AI, developed for space missions, enhance industries on Earth. Satellite technology, born out of the necessity to communicate across vast distances, now underpins the global telecommunications framework. In this way, commercial and international ventures in space exploration showcase how progress in one domain can ripple out to effect change on a global scale.

Furthermore, the educational and inspirational impacts of these ventures are immeasurable. Space exploration serves as a potent catalyst for STEM (Science, Technology, Engineering, and Mathematics) education, inspiring the next generation of scientists, engineers, and dreamers. The collaborative achievements in space demonstrate the tangible results of pursuing a career in these fields, igniting curiosity and ambition in young minds across the globe. Projects like the global Mars Rover Challenge invite students to design and operate rover missions, offering hands-on experience in this exciting arena.

Yet, despite these lofty aspirations and achievements, the realm of commercial and international space ventures faces challenges. Issues of space debris, sustainable exploration, and the militarization of space must be addressed. The Kessler Syndrome, a scenario where the density of objects in low Earth orbit is high enough to cause collisions that generate more space debris, could jeopardize future missions. International cooperation in mitigating debris and fostering sustainable practices is imperative for the long-term viability of space exploration.

The road to becoming a multi-planetary species is fraught with technical, economic, and ethical hurdles. Nevertheless, the tapestry of commercial and international ventures weaves a promising narrative. It's a story where ambition, collaboration, and innovation are the threads that entwine to create a vision of the future that transcends the limitations of our terrestrial abode.

As private companies carry forward the torch of human endeavor, and as nations come together in the spirit of curiosity and exploration, the potential for what lies ahead seems boundless. From space tourism to asteroid mining, from lunar bases to interplanetary travel, these commercial and international undertakings are writing the next chapters in the annals of human achievement. These ventures are more than just missions; they are testament to our unyielding desire to push the boundaries of what is possible, to boldly explore the unknown, and to seek out our place in the infinite expanse of the cosmos.

CHAPTER 22:
ROBOTICS

From the automata of ancient Greece to the sophisticated machines of today, robotics has charted an extraordinary course through human history. Early robots, often little more than mechanical curiosities, gradually evolved into vital components of industrial automation, revolutionizing manufacturing and production. Today, robots transcend factory floors, finding roles in healthcare, agriculture, domestic environments, and even space exploration. They assist surgeons with precision, harvest crops with efficiency, clean our homes, and traverse alien landscapes. This evolution underscores not only our relentless pursuit of innovation but also our deep-seated desire to enhance, simplify, and sometimes even transcend the limits of human capability. Robotics stands as a testament to how far we've come and hints at a future filled with boundless possibilities and uncharted potentials.

Early Robots and Automation

As we edge closer to the cutting-edge robotics of today, it helps to look back at the origins of robots and automation. From the mythical creations of ancient civilizations to the clockwork wonders of the Middle Ages, the idea of machines performing tasks autonomously is far from new. These early concepts not only showcase human ingenuity but also reflect our perennial desire to push boundaries and redefine possibilities.

The genesis of automated devices can be traced as far back as ancient Greece. The Greek engineer Hero of Alexandria, who was active around the first century AD, created several automated devices, many of which seem incredibly advanced even by today's standards. One of his notable inventions was the 'aeolipile,' a simple steam engine that operated on principles strikingly similar to modern jet propulsion. Hero's inventions also included automated theaters and even a vending machine that dispensed holy water. These creations, though rudimentary by contemporary measures, laid the foundational concepts for mechanical automation.

Fast forward a few centuries, and the concept of automation continued to evolve. In medieval Europe, intricate clockwork mechanisms began to capture the imagination. One significant example was the astronomical clock in Prague, built in the 15th century and still operational today. This clock wasn't just a timepiece; it was a marvel of mechanical engineering, featuring rotating figures, astronomical dials, and even a rooster that crows at the hour. Such clocks represented mankind's burgeoning desire to merge artistry with technology.

By the Renaissance, artists and engineers like Leonardo da Vinci were designing some of the earliest forms of robots. Leonardo's sketches include a mechanical knight, capable of limited movement and gestures. Although it never moved beyond the concept phase, Leonardo's work was a powerful testament to the possibilities of combining human-like forms with mechanical systems.

The onset of the Industrial Revolution in the 18th century marked a pivotal shift. This era brought about the invention of more complex and practical automated systems, primarily designed to improve efficiency and productivity in industries. The development of textile machinery, notably the spinning jenny and the power loom, transformed how goods were produced. These machines could operate continuously and with far greater precision than human workers, thereby setting the stage for mass production.

Charles Babbage, often regarded as the father of the computer, also contributed to early automation concepts. His designs for the Analytical Engine in the 1830s envisioned a programmable machine that could perform calculations and process data. While his machine was never completed, Babbage's ideas significantly influenced the development of modern computing and automated processes.

By the late 19th and early 20th centuries, the concept of automation took a giant leap with the invention of assembly lines. Most famously implemented by Henry Ford in his automobile factories, the assembly line used continuous workflows, with workers and machines performing specialized tasks in sequence. This not only revolutionized production capabilities but also significantly lowered costs, making products like cars accessible to a broader audience.

World War II further accelerated developments in automation and robotics, driving rapid advancements in technology. The need for efficient manufacturing and complex machinery spurred innovations such as automated turrets and bombers, laying the groundwork for post-war technological advancements. It was during this period that the term 'robot' gained prominence, derived from the Czech word 'robota,' meaning forced labor. Initially introduced in Karel Capek's 1920 play "R.U.R. (Rossum's Universal Robots)," robots depicted mechanical beings that could perform human tasks, setting a cultural precedent that endures to this day.

As war gave way to peacetime, the ensuing technological boom saw significant strides in robotics and automation. The 1950s and 1960s were the golden era of industrial automation. Factories increasingly adopted automated machinery to increase efficiency, while researchers and engineers began to explore the potential of robots in various fields. This period also saw the introduction of the first programmable robot,

"Unimate," designed by George Devol and Joseph Engelberger. Unimate was initially employed in General Motors' factory to perform tasks like welding and die-casting.

Meanwhile, the concept of cybernetics started to gain popularity. Pioneered by mathematician Norbert Wiener, cybernetics revolves around the study of communication and control in living organisms and machines. Wiener's work highlighted the parallels between biological systems and mechanical systems, reinforcing the idea that machines could be built to mimic life processes. This thinking laid the intellectual foundation for more advanced robotics and automation.

Another critical milestone was achieved with the advent of computer numerical control (CNC) machines. CNC machines, which began gaining traction in the 1960s, combined computer technology with machine tools to perform highly precise and customizable tasks. This development revolutionized industries like aerospace, automotive, and electronics, where precision and repeatability are crucial.

The proliferation of microprocessors in the late 20th century further transformed the landscape. These tiny yet powerful computing units enabled the creation of more sophisticated robots and automated systems. Mobile robots began to emerge, capable of navigating diverse environments and performing a range of tasks. The convergence of microelectronics, information theory, and mechanical engineering led to the birth of modern robotics.

Early robots and automation have had an indelible impact on contemporary life and industry. The mechanical marvels of ancient Greece, the intricate clocks of medieval Europe, and the automated factories of the Industrial Revolution all reflect humanity's unrelenting quest to push the boundaries of technological capability. They laid the groundwork for the sophisticated autonomous systems we encounter today, from robotic surgeons to automated warehouses.

As we consider these early robots and their creators, we can appreciate the essence of innovation—an imaginative spark that propels us forward. These early efforts provide a valuable perspective on our current technological advancements and remind us of a timeless truth: the drive to automate, innovate, and revolutionize our world is as old as civilization itself. With each new invention, we not only solve immediate problems but also pave the way for future breakthroughs, creating a continuously evolving tapestry of human ingenuity.

Modern Robotics in Daily Life

The integration of robotics into our daily lives is no longer a notion confined to science fiction. These technological marvels have seamlessly woven themselves into the very fabric of our day-to-day activities, making them indispensable in many ways. Whether it's through the household Roomba vacuum navigating around living room furniture or the sophisticated robotic arms in manufacturing plants, modern robotics is enhancing efficiency, precision, and convenience across various facets of life.

One of the most visible applications of robotics in households is through robotic vacuum cleaners. These intelligent devices use a combination of sensors, cameras, and algorithms to map out rooms and efficiently clean floors while avoiding obstacles. They offer not just convenience but also a glimpse into how autonomous systems can adapt to a dynamic human environment.

However, the household is just the beginning. Modern robotics has found its way into the world of personal care, where robots assist the elderly and differently-abled individuals. From robotic exoskeletons that help paraplegics walk to companion robots providing emotional support, these machines are transforming lives in ways previously unimaginable. Robotics in healthcare settings is also noteworthy, with robots assisting in surgeries, ensuring precision that even the most skilled human hands might not achieve. Robots like the Da Vinci Surgical System allow surgeons to perform minimally invasive surgeries with unmatched accuracy.

Education is another sphere where robotics is making significant strides. Educational robots such as NAO and Pepper are being used in classrooms to teach subjects ranging from coding to social skills. These robots can interact with students, answer questions, and even adapt teaching methods to suit individual learning styles. Such interactions can make learning more engaging and effective, preparing the younger generation for a future where technology plays an even more significant role.

Retail and hospitality sectors are not far behind in adopting robotic technology. In some stores, robots are being used to stock shelves, manage inventory, and even greet customers. This shift not only enhances operational efficiency but also reduces the need for human intervention in mundane tasks, allowing staff to focus on more complex and customer-interactive roles. In hotels, robots are increasingly being used for room service, concierge tasks, and housekeeping, providing a unique experience for guests while streamlining operations.

In agriculture, robotics is revolutionizing how we approach the cultivation and harvesting of crops. Agricultural robots, or agrobots, can perform tasks such as seeding, weeding, and harvesting with precision. These robots use GPS technology, machine vision, and even AI algorithms to optimize agricultural practices, ensure sustainability, and increase yield. The result is a more efficient food production system capable of meeting the growing demands of the global population.

Transportation also benefits immensely from modern robotics. Autonomous vehicles, including self-driving cars and drones, represent the future of transportation. These technologies promise to reduce traffic congestion, lower accident rates, and provide greater accessibility for those unable to drive. Companies like Tesla, Waymo, and Uber are at the forefront of developing autonomous driving technology, while drone delivery services are being tested and implemented by giants like Amazon and UPS.

In the realm of home automation, robots act as central figures. Smart home assistants such as Amazon's Alexa, Google Home, and Apple's Siri may not have the physical form of traditional robots, but they perform a myriad of automated tasks. From controlling lighting and temperature to managing home security systems, these intelligent assistants bring a level of automated convenience that's becoming increasingly prevalent.

The entertainment industry is also leveraging robotics to create captivating experiences. Animatronics and robotic characters in theme parks like Disney World add a level of realism and interactivity that enhances visitor experiences. Robots are also used in the film industry for special effects, bringing scenes to life in ways that CGI can't always achieve. Additionally, robotic musicians and artists are exploring new frontiers in creative expression, challenging our perceptions of art and creativity.

It's crucial to recognize the role of robotics in fostering inclusivity and independence. Assistive robots are being developed to help individuals with disabilities perform daily tasks that might otherwise be challenging. From robotic arms that can aid in eating to autonomous wheelchair systems that navigate complex environments, these innovations empower individuals and offer a better quality of life.

Modern robotics is also leaving its mark on the world of finance and banking. The use of robotic process automation (RPA) helps in performing repetitive and routine tasks with high accuracy and speed. Banks and financial institutions are utilizing RPA to handle tasks such as data entry, transaction processing, and regulatory compliance. As a result, human employees are freed to focus on more intricate tasks that require analytical thinking and creativity. Additionally, customer service bots in banking are providing round-the-clock assistance to clients, enhancing the overall customer experience.

In the culinary world, robots are starting to influence how we prepare and serve food. Automated kitchen systems can cook complex dishes with precision and consistency, ensuring high-quality output every time. Restaurants are adopting robotic chefs and waitstaff to streamline operations and offer unique dining experiences. These robotic systems can handle everything from food preparation and cooking to plating and serving, transforming the traditional kitchen landscape.

As we forge ahead, the ethical considerations surrounding robotics become increasingly urgent. The balance between human jobs and robotic automation remains a contentious issue. While robots can perform many tasks with greater efficiency and less error, the implications for the workforce are profound. There's an ongoing conversation about how to ensure that the rise of robotics doesn't lead to widespread job displacement but rather creates new opportunities for human-robot collaboration.

Moreover, the concept of human-robot interaction (HRI) is critical in developing safe and effective robotic systems. Researchers and developers are focusing on how robots can better understand and predict human behavior to interact more naturally and safely. This involves sophisticated programming, machine learning, and a deep understanding of human psychology. The goal is to create robots that are not just tools but collaborators that understand and respond to human needs and emotions.

Despite these challenges, the potential benefits of robotics in daily life are vast. As technology advances, robots will become even more integrated into our personal and professional lives, offering solutions to complex problems and enhancing our overall lifestyle. From healthcare and education to entertainment and everyday household chores, the presence of robots in our world has only just begun to shape the contours of our future.

Ultimately, the essence of modern robotics in daily life lies in its ability to augment human capabilities and improve the quality of life. Robotics is not just about replacing human effort but about complementing it, offering precision, efficiency, and the ability to perform tasks beyond human capability. It's a partnership between human ingenuity and technological advancement, and as we continue to innovate, this partnership will only grow stronger, guiding us toward a future where robots and humans thrive together.

CHAPTER 23:
VIRTUAL AND AUGMENTED REALITY

Virtual and augmented reality have metamorphosed from being mere simulations to immersive experiences that deeply engage the senses and elevate our interaction with the digital realm. Once the exclusive domain of sci-fi dreams and niche tech enthusiasts, these innovations now infiltrate diverse spheres such as gaming, medicine, education, and beyond. Imagine surgeons practicing intricate procedures in a risk-free digital world, or students exploring ancient civilizations in a classroom transformed by AR overlays. This technology's potential to merge the physical and digital worlds inspires a reevaluation of what is possible, underscoring a future where our realities are only limited by our imaginations.

From Simulations to Immersive Experiences

The realm of virtual and augmented reality stands as a testament to humanity's insatiable curiosity and its quest to push the boundaries of perception. From the rudimentary simulations of the past, wherein pixelated environments served as mere shadows of reality, we've vaulted into an era of breathtaking immersive experiences. This evolution didn't happen overnight but is the culmination of decades of technological advancements and imaginative forethought.

In the early days, simulations primarily served functional purposes. Military training programs and flight simulators were among the first to harness basic virtual reality (VR). These primitive systems were groundbreaking for their time, offering a safer and more controlled environment for training. It was during this period that the potential of VR started to become apparent, highlighting the possibilities of creating alternate realities for various applications.

As time went on, computing power increased, graphics improved, and the technology became more accessible. The leap from curtained-off arcade machines to household gaming consoles brought the virtual world into our living rooms. Who could forget the first time they donned a VR headset and found themselves immersed in a cartoonish but captivating new world? Each new iteration expanded the boundaries, inching closer to today's hyper-realistic experiences.

While VR has made significant strides, augmented reality (AR) has followed a parallel yet distinct path. Unlike VR, which creates entirely synthetic environments, AR overlays digital information onto the physical world. Early iterations of AR were limited to heads-up displays in fighter jets, showing essential data without distracting the pilot. Nowadays, AR applications, like those on our smartphones, can transform our understanding and interaction with the world, from navigation aids to immersive gaming experiences.

Gaming, undoubtedly, has been a powerful driving force behind the advancements in both VR and AR. The lure of a fully immersive gaming experience has spurred massive investments and rapid technological progress. Titles that were once only the dream of science fiction are now available for anyone with a console or a PC powerful enough to run them. This trend has only accelerated with the advent of social VR platforms, allowing people from all corners of the globe to meet and interact in fantastical worlds.

Beyond entertainment, these technologies have found applications in education, healthcare, and industry. Virtual field trips can transport students to Ancient Rome, while medical professionals use VR for intricate surgical simulations. In industrial design, AR overlays can provide real-time data visualization, streamlining workflows and reducing errors. These are but a few examples of how VR and AR enrich our lives and enhance our capabilities.

The technological landscape of VR and AR also poses tantalizing philosophical questions about the nature of reality itself. At what point does a simulated experience become indistinguishable from reality? What ethical considerations arise when creating fully immersive experiences that can profoundly impact the psyche? These are critical questions as we stand on the cusp of further revolutionary advancements.

Moreover, the rapid pace of development in VR and AR technology challenges our perceptions and opens new avenues for creativity. Artists and storytellers are no longer constrained by the physical world; they can build entire universes limited only by their imagination. This capability redefines traditional narratives and offers unprecedented ways to tell stories and share experiences.

The hardware behind these technologies has also evolved dramatically. Early VR systems were bulky, tethered, and often caused motion sickness. Today's headsets are sleek, wireless, and come equipped with advanced tracking systems that map out the user's environment, reducing the likelihood of disorientation. Similarly, AR has moved from cumbersome head-mounted displays to lightweight glasses and mobile apps, making it more practical and user-friendly.

The future trajectory of VR and AR is set to revolutionize even more aspects of our daily lives. Imagine virtual business meetings in photorealistic settings, or AR providing hands-free information for tasks as mundane as cooking or as complex as engineering. The convergence of these technologies with artificial intelligence and the Internet of Things is poised to transform our interaction with the digital and physical worlds.

In this brave new era, the line between the real and the virtual may blur, but one thing remains clear: the journey from rudimentary simulations to immersive experiences represents a pivotal leap in human technological evolution. As we continue to explore and expand these boundaries, the possibilities are boundless, promising a future limited only by our collective innovation and creativity.

Applications in Various Fields

Virtual and Augmented Reality (VR and AR) are not merely novelties confined to gaming or entertainment. These transformative technologies have found profound applications across a broad spectrum of fields, reshaping the way we perceive and interact with the world around us.

In the realm of education, VR and AR have opened doors to immersive learning experiences that were once unimaginable. Imagine stepping into a history class and being transported to ancient Rome, standing amidst the grandeur of the Colosseum while listening to a lecture. With VR, this level of interaction isn't just possible; it's already happening. Similarly, AR can superimpose educational content onto physical surroundings, making abstract concepts come to life in a tangible manner. This interactive form of learning isn't just engaging; it's proving to be more effective, as students are able to retain information better when they experience it firsthand.

Healthcare is another field where VR and AR have made significant strides. Surgeons are now utilizing VR to simulate complex surgeries, allowing them to practice procedures without any risk to patients. This not only enhances their skills but also leads to better patient outcomes. AR, on the other hand, can assist during live surgeries by overlaying vital information and guides directly onto the surgeon's field of view, reducing the likelihood of errors.

In mental health therapy, VR is being used to help patients confront and manage issues like PTSD and phobias in controlled, virtual settings. Creating these safe spaces enables individuals to face their triggers gradually and constructively, with a therapist guiding them through each step. The potential here is vast, offering new hope and methods for treatment in a field that desperately needs them.

In the architecture and real estate industries, VR and AR are revolutionizing how spaces are designed, sold, and interacted with. Architects can now create detailed virtual models of their projects, allowing clients to take virtual walkthroughs before a single brick is laid. This not only aids in visualization but also in identifying potential design flaws, saving both time and resources. Real estate agents are using similar technology to show homes to potential buyers from halfway across the world, making the market more accessible and transparent than ever before.

The retail sector is also embracing VR and AR to enhance customer experience. Imagine trying on clothes in a virtual dressing room or seeing how a piece of furniture would look in your home without having to actually bring it there. AR apps are making shopping more interactive and personalized, leading to higher customer satisfaction and reduced return rates.

If we shift our focus to the automotive industry, VR is playing a crucial role in both design and training. Car manufacturers use virtual prototypes to test new models, allowing them to identify and rectify issues long before the physical models are built. This not only speeds up the development process but also ensures higher quality and performance. On the consumer side, potential buyers can now take virtual test drives, offering a taste of the driving experience without setting foot in a showroom. Additionally, AR is being integrated into driver assistance systems, providing real-time information and enhancing navigation through head-up displays.

The field of tourism has also seen a significant boost from VR and AR technologies. Virtual travel experiences are allowing individuals to explore exotic destinations from the comfort of their homes. These experiences can spark a desire to travel and provide preliminary information about a location before actually visiting. Meanwhile, AR-enabled guide apps enrich real-world travel by providing historical and contextual information about landmarks, making the journey more informative and engaging.

In manufacturing, both VR and AR are being harnessed to revolutionize the production process. VR can simulate assembly lines and production flows, optimizing efficiency and identifying potential bottlenecks without disrupting actual operations. AR is being used on the factory floor for training purposes and real-time problem-solving. Workers equipped with AR glasses can receive step-by-step assembly instructions or maintenance guides, significantly reducing errors and improving productivity.

Even in the military, VR and AR have practical applications. VR is being used for combat training, allowing soldiers to engage in simulated environments that mimic real-world scenarios without the associated risks. This kind of training can be repeated endlessly, sharpening skills and preparing troops more effectively. AR, meanwhile, can enhance situational awareness in the field by overlaying critical data and information onto a soldier's visor, providing strategic advantages in real-time.

One often overlooked area where VR and AR are making inroads is in environmental science and conservation. By creating immersive experiences, these technologies can raise awareness and educate people about environmental issues in a compelling way. For example, a VR experience could take someone underwater to witness coral bleaching, driving home the urgency of climate change in a way that no documentary could match. AR, too, can assist in fieldwork by providing real-time data overlays, helping researchers gather and analyze ecological data more efficiently.

The sports industry isn't far behind either. Teams are using VR to train athletes by simulating game scenarios, improving reaction times, and strategy planning. Fans, too, get to enjoy immersive experiences; VR allows for virtual courtside seats that offer angles and perspectives that even the best TV broadcast can't match. AR is also being used in stadiums to enhance the live viewing experience by providing real-time stats and player information directly to one's smartphone or AR glasses.

Furthermore, the field of arts and entertainment continues to be revolutionized by VR and AR. Museums and galleries are incorporating AR to provide enriched, interactive experiences for visitors, offering layers of information and context that traditional displays couldn't. VR is opening up new frontiers in filmmaking and storytelling, offering 360-degree narratives that place the audience in the heart of the action. These applications are not just innovative; they're transforming how culture and knowledge are consumed and appreciated.

In journalism and media, VR is adding a new dimension to storytelling. Reporters can now create immersive reports that allow viewers to experience news events firsthand. Whether it's the aftermath of a natural disaster or a bustling marketplace in a distant country, VR enables a level of empathy and understanding that traditional media simply can't match. AR is enhancing the way we consume everyday media as well, with interactive articles and augmented reality graphics that provide deeper engagement with content.

Commercial applications are continuously evolving as well, including exciting uses in marketing and advertising. Brands are leveraging VR and AR to create more memorable and engaging advertisements. Virtual showrooms, AR product demos, and immersive brand experiences are just the tip of the iceberg. These technologies offer compelling ways for companies to connect with consumers, creating lasting impressions and fostering a deeper connection with their brands.

To sum up, the adaptive and immersive qualities of Virtual and Augmented Reality are providing groundbreaking solutions across a multitude of fields. Their applications go far beyond entertainment, filling critical roles in education, healthcare, real estate, automotive, tourism, manufacturing, military, environmental science, sports, arts and entertainment, journalism, and commercial sectors, among many others. Instead of merely watching technology unfold, various industries are now actively shaping their futures through VR and AR, fundamentally altering what we perceive as possible. The extensive adoption and innovative applications of these technologies underscore their pivotal role in continuing the human journey toward progress and evolution.

CHAPTER 24:
ARTIFICIAL INTELLIGENCE

A rtificial Intelligence, often abbreviated as AI, represents one of the most transformative leaps in our technological journey, impacting every facet of modern life. Emerging from the foundation of early research in the 20th century, AI has rapidly evolved from theoretical concepts to practical applications that permeate our daily routines. Its range extends from simple algorithms powering search engines to complex systems behind autonomous vehicles and advanced medical diagnostics. Through machine learning and neural networks, AI systems have learned to mimic cognitive functions such as learning, reasoning, and problem-solving, thereby acting as both our assistants and challengers in numerous domains. The era of AI offers unparalleled opportunities for innovation and efficiency but also poses profound questions about ethics, privacy, and the future of human labor. As we delve into this era, we're standing at the crossroads of boundless potential and significant responsibility, making AI not just a tool but a pivotal chapter in human progress.

Foundations and Early Research

The journey of artificial intelligence (AI) didn't begin with the sophisticated algorithms or machine learning models we know today. It traces back to philosophical ponderings on what it means to think and learn, a topic that intrigued ancient Greek philosophers. They posited that human intelligence could possess elements replicable by machines. Ideas about automata in mythologies and automaton craftsmanship in the Hellenistic period also signify an early interest in replicating human functions through machinery.

The real pivot toward AI's foundational research commenced in the 20th century. It was a time marked by burgeoning curiosity about computation, inspired by the works of visionaries like Alan Turing and John von Neumann. Turing's groundbreaking paper "Computing Machinery and Intelligence," published in 1950, posed the poignant question: "Can machines think?" This work introduced the concept of the Turing Test, a criterion measuring machines' ability to exhibit intelligent behavior indistinguishable from humans.

Meanwhile, John von Neumann contributed through his development of the architecture that underpins most computer systems today. He introduced the stored-program concept, which allowed machines to be more flexible and powerful. Von Neumann's work laid the bedrock for complex computational operations central to AI research.

In the same vein, the 1956 Dartmouth Conference is often credited as the birth of artificial intelligence as an academic discipline. Hosted by John McCarthy, Marvin Minsky, Nathaniel Rochester, and Claude Shannon, the conference gathered pioneers in the field to brainstorm on the possibilities of simulating aspects of human intelligence. This assembly marked the coinage of the term "artificial intelligence" and catalyzed numerous research projects that laid the groundwork for future advancements.

Creating machines that could learn from data and experiences became the next major milestone. Initial attempts include the Perceptron, an early neural network model proposed by Frank Rosenblatt in 1958. Although the Perceptron was limited in capabilities, it provided critical insights into machine learning's potential and constraints. It essentially modeled the simplest neural pathways, representing an embryonic step toward the elaborate neural networks we see today.

As foundational research expanded, the 1960s and 1970s saw a surge in enthusiasm and funding for AI. This period presented a bouquet of hopes and, sometimes, unrealistic expectations. Key players like Herbert Simon and Allen Newell worked on Logic Theorist and General Problem Solver, some of the earliest AI programs designed to mimic human problem-solving. Parallel efforts, such as those by Edward Feigenbaum and colleagues, led to the creation of DENDRAL, the first expert system, initially designed for chemical analysis. This leap from theory to application revealed AI's potential across various fields.

The field was not without its setbacks. During the 1970s and 1980s, AI research faced the so-called "AI winter," a period of reduced funding and interest. Early systems often struggled with scalability and the computational limitations of the time. The initial hype gave way to pragmatism as critics highlighted the widening gap between aspirations and actual capabilities.

Despite these challenges, substantial theoretical progress continued. The development of backpropagation algorithms in the mid-1980s revived neural networks research. This era also witnessed increased interest in probabilistic reasoning and logic programming, which spurred innovations in natural language processing, computer vision, and robotics.

The academic community began to appreciate the intricacies involved in mimicking human intelligence. Efforts diversified into subfields like knowledge representation, linguistic structures, and cognitive modeling. Researchers became more interdisciplinary, drawing insights from psychology, neuroscience, and linguistics to inform algorithmic development. The works of Noam Chomsky on syntax and grammar structures, for instance, influenced natural language processing endeavors.

In the 1990s, the advent of more advanced computing hardware and the expansion of the internet contributed to AI's resurgence. Machine learning, in particular, flourished with the increased availability of data and enhanced computational power. Algorithms like Support Vector Machines and the aforementioned backpropagation for neural networks provided powerful tools for solving complex problems. AI research also began to gain commercial traction, with early applications in data mining, customer service automation, and financial forecasting.

It was becoming clear that AI was not just a theoretical pursuit but a practical technology capable of transforming industries. Initial applications were seen in areas such as autonomous systems, with self-driving cars making their debut in research labs. Despite facing new challenges, the foundational research during this period built a robust base for AI's exponential growth in the 21st century.

These early years of AI research laid a comprehensive foundation for modern advancements. They involved a unique blend of mathematical rigor, computational innovation, and philosophical inquiry, each contributing to our current understanding of artificial intelligence. Researchers faced numerous challenges but also set the path for breakthroughs yet to come.

AI in Everyday Life

Artificial Intelligence (AI) is no longer a distant, futuristic concept; it's very much a part of our daily routines and tasks. From the way we communicate to the services we use, AI has seamlessly integrated into our lives. Consider the virtual assistants like Alexa, Siri, or Google Assistant. These digital helpers can set reminders, control smart home devices, answer questions, and even engage in small talk. This once sci-fi notion of speaking to a machine has become second nature.

Another area where AI has made remarkable strides is in the personalization of online experiences. Streaming services like Netflix and

Spotify use AI algorithms to suggest movies, shows, and music based on our past behavior and preferences. This kind of personalization also extends to online shopping. Retail giants like Amazon deploy sophisticated AI systems to recommend products, making shopping more efficient and tailored to individual needs.

The realm of social media offers another compelling example of AI's ubiquitous presence. AI algorithms curate our news feeds, recommend friends or accounts to follow, and even flag inappropriate content. It's this unseen hand that ensures you see posts from close friends at the top of your feed or get targeted advertisements based on your browsing history. The algorithms are ever-evolving, learning from each of our interactions to create a more compelling user experience.

AI is also revolutionizing healthcare in ways that we might not immediately notice. Machine learning models are being used to predict diseases, analyze medical images, and even suggest personalized treatment plans. Wearable devices equipped with AI capabilities can monitor vital signs in real-time, providing critical data to healthcare providers and enabling timely interventions. It's not far-fetched to imagine a future where AI can detect signs of disease before we even experience symptoms.

In the financial sector, AI is transforming how we manage our money. Financial institutions use AI to detect fraudulent activities, provide customer service via chatbots, and even offer investment advice. Robo-advisors are an innovation born out of AI that can help create and manage investment portfolios tailored to individual risk appetites and financial goals. With AI, the complexities of financial planning become much more accessible to the average person.

The automobile industry offers yet another fascinating glimpse into AI's real-world applications. Autonomous vehicles, powered by sophisticated AI systems, promise to revolutionize how we think about transportation. While fully self-driving cars are still a work in progress, many vehicles today already come equipped with advanced driver-assistance systems (ADAS). These systems can perform tasks such as adaptive cruise control, lane-keeping assistance, and even automated parking.

Entertainment isn't left out either. AI is increasingly being used in the creation of art, music, and literature. AI algorithms can compose original pieces of music, write poetry, and even create visual art. The creative potential of AI is immense, opening up new avenues for artists and creators. At the intersection of technology and creativity, AI is pushing the boundaries of what's possible, and the results are often astonishing.

In education, AI tools are making learning more interactive and personalized. Adaptive learning platforms adjust the difficulty of tasks based on student performance in real-time, ensuring that individuals learn at a pace that suits them best. Virtual tutors powered by AI can provide extra help on demand, breaking down complex topics into more digestible parts. This transformation is particularly beneficial for remote learning environments, which have become more prevalent due to global circumstances.

The smart home ecosystem is another domain where AI shines brightly. AI-powered devices, from thermostats that learn your temperature preferences to smart refrigerators that can suggest recipes based on available ingredients, make home management simpler and more efficient. Voice-controlled lighting, security systems, and even robotic vacuum cleaners bring a new level of convenience and automation to everyday tasks.

AI is also playing a critical role in environmental conservation efforts. By analyzing data collected from satellites, drones, and ground sensors, AI can help us understand and predict environmental changes with unprecedented accuracy. These insights are crucial for making informed decisions about conservation strategies, resource management, and impact reduction. Furthermore, AI-driven models can simulate climate scenarios, offering valuable projections that guide policy decisions at both local and global levels.

Customer service is another area undergoing a quiet yet significant transformation due to AI. Many companies now use AI-driven chatbots to handle customer inquiries, provide support, and even make product recommendations. These bots can handle multiple queries simultaneously and are available around the clock, ensuring that customers receive timely and accurate assistance. The use of AI in customer service frees up human agents to handle more complex issues, enhancing overall efficiency and customer satisfaction.

AI's role in enhancing accessibility for individuals with disabilities cannot be overstated. For example, real-time transcription services powered by AI can help those with hearing impairments by converting spoken words into text instantly. Similarly, computer vision technology enables applications that assist visually impaired individuals by describing their surroundings or reading text out loud. These advancements empower people with disabilities to navigate the world with greater ease and independence.

In the world of sports, AI is making an impact by providing performance analytics and injury prevention strategies. AI-driven software can analyze an athlete's movements, identifying inefficiencies or potential risk factors. This level of analysis enables coaches and athletes to make data-driven decisions, optimizing training regimes and reducing the likelihood of injury. AI's predictive capabilities help create safer and more effective training environments.

Finally, AI is proving indispensable in enhancing public safety and security. Surveillance systems equipped with AI can detect unusual activities and alert authorities in real-time, significantly reducing response times. Predictive policing models use historical data and machine learning algorithms to identify and anticipate crime hotspots, enabling law enforcement agencies to allocate resources more effectively. While these applications raise important ethical considerations, their potential to improve public safety is undeniable.

The integration of AI into our daily lives represents a significant transformative force, ushering in an era of unprecedented convenience, efficiency, and innovation. As AI continues to evolve, its applications will undoubtedly expand, permeating even more aspects of our lives. The key to harnessing the full potential of AI lies in addressing ethical concerns, ensuring data privacy, and creating inclusive technologies that benefit all segments of society. The future of AI in everyday life promises to be both exciting and challenging, offering opportunities to reimagine how we live, work, and interact with the world around us.

CHAPTER 25:
THE FUTURE OF TECHNOLOGY

The march of technology has been relentless, yet the future holds promises that even the greatest minds of our past could scarcely imagine. In the coming years, innovations like quantum computing and advanced AI could redefine our very understanding of reality. Future societies may grapple with profound ethical choices as technology integrates further into our daily lives, from AI-driven healthcare that can predict illnesses before they manifest, to sustainable energy solutions that might finally curb climate change. Technologies once confined to the realms of science fiction, such as human-robot symbiosis and brain-computer interfaces, are steadily becoming our new reality, challenging us to rethink the essence of being human. The horizon is brimming with both potential and responsibility, urging us to navigate this coming era with wisdom and foresight.

Predicting Future Innovations

As we stand on the precipice of the future, predicting what lies ahead in the technological realm is both thrilling and daunting. It's a task that involves extrapolating current trends, imagining radical breakthroughs, and understanding the evolving needs and aspirations of humanity. History shows us that while some predictions have been astonishingly accurate, others have veered off course. The flying cars and lunar habitats many envisioned decades ago haven't quite materialized, but instead, we've seen the rise of smartphones, the internet, and artificial intelligence—all innovations that have profoundly transformed our world.

Looking ahead, one of the most promising domains is artificial intelligence (AI). We can expect AI to integrate even more seamlessly into our daily lives, enhancing everything from healthcare diagnostics to personalized learning. AI systems will likely become more autonomous, requiring less human intervention and making decisions based on vast datasets. Imagine AI doctors that can diagnose diseases with unprecedented accuracy by analyzing millions of medical records, or tutors that adapt their teaching methods to fit each student's unique learning style. The potential here is vast, but it comes with responsibilities and ethical considerations, particularly concerning privacy, bias, and job displacement.

The internet itself is another area ripe for transformation. With the advent of 5G and beyond, connectivity will become ubiquitous and instantaneous. This could pave the way for more immersive virtual and augmented reality experiences, where the lines between the physical and digital worlds blur even further. Virtual meetings could feel like face-to-face interactions, and augmented reality could enhance everything from navigation to social interactions. Furthermore, the Internet of Things (IoT) will continue to interconnect devices, making our homes, cities, and transportation systems smarter and more efficient.

One can't discuss the future of technology without mentioning renewable energy. As the world grapples with the dire consequences of climate change, innovations in this area are not just expected—they're essential. Solar and wind technologies are rapidly advancing, becoming more efficient and less costly. In the near future, we may see breakthroughs in energy storage, such as advanced batteries that can store solar power for use at night or during cloudy days. There's also a growing interest in exploring alternative energy sources, such as fusion power, which could provide virtually limitless energy if the technical challenges can be overcome.

Space exploration is another frontier where we can anticipate groundbreaking advancements. The commercialization of spaceflight, spearheaded by companies like SpaceX and Blue Origin, promises to make space more accessible. Missions to Mars and beyond will likely become a reality within our lifetimes, leading to potential human settlements on other planets. These endeavors will require new technologies in life support, habitat construction, and space travel, pushing the boundaries of what we currently deem possible.

Biotechnology is poised to make significant strides as well. With tools like CRISPR and other gene-editing technologies, we may soon have the ability to eradicate genetic diseases, enhance human capabilities, and even extend lifespan. The ethical implications of such technologies are profound and must be carefully considered, but the possibilities for improving human health and wellbeing are immense. Personalized medicine, tailored to an individual's genetic makeup, could become the norm, leading to more effective and targeted treatments.

In the realm of transportation, autonomous vehicles are on the brink of revolutionizing how we move. Self-driving cars and trucks could reduce traffic accidents, decrease congestion, and make transportation more efficient. Moreover, advancements in electric vehicles, coupled with improvements in battery technology, will likely lead to a cleaner and more sustainable transport system. Hyperloop technology, which aims to transport people and goods at near-supersonic speeds through vacuum tubes, could redefine long-distance travel, making it faster and more eco-friendly.

The advent of quantum computing is another development that could reshape numerous fields. Unlike traditional computers, which process information in binary, quantum computers use qubits that can exist in multiple states simultaneously. This allows them to solve complex problems much faster than classical computers. The potential applications are vast, ranging from cryptography and materials science to drug discovery and climate modeling. Quantum computing might help us tackle problems previously deemed insurmountable.

Another pivotal area of future innovation is in the domain of digital ethics and cybersecurity. As our dependence on digital systems grows, so does the need to protect them from malicious attacks. The future will likely see the development of more sophisticated cybersecurity measures, harnessing AI to predict and neutralize threats before they can cause harm. Moreover, the ethical implications of emerging technologies, such as data privacy and the digital divide, will need to be addressed to ensure an inclusive and fair technological future.

In a world where technology is advancing at breakneck speed, the role of education and lifelong learning becomes crucial. As new technologies emerge, the skills required in the workforce are constantly evolving. Future innovations will likely make education more personalized and accessible. AI-driven learning platforms can tailor educational content to suit individual learning paces and styles, making education more effective. Additionally, virtual and augmented reality could provide immersive learning experiences, allowing students to explore historical events, conduct virtual science experiments, or even practice complex surgical procedures.

Moreover, the concept of smart cities is becoming increasingly relevant as urbanization continues to rise. These cities leverage technology to improve the quality of life for their residents, making urban environments more efficient, sustainable, and livable. Innovations in this area could include smart grids for energy distribution, intelligent traffic management systems, and advanced waste management technologies. The integration of IoT devices throughout the urban landscape promises to make our cities not only more responsive to the needs of their inhabitants but also more resilient in the face of challenges such as climate change and population growth.

Finally, as we ponder the future of technology, it's essential to consider the role of collaboration and interdisciplinary research. Many of the most significant technological advancements have emerged from the intersection of different fields. For instance, advancements in materials science combined with AI could lead to the development of new, more efficient drug delivery systems. Similarly, the fusion of biology and engineering is paving the way for biohybrid robots that could have applications ranging from medical procedures to environmental monitoring.

In conclusion, while predicting future innovations is an inherently uncertain endeavor, the trends and emerging technologies discussed provide a glimpse into the possibilities that lie ahead. The future of technology promises to be filled with remarkable advancements that will transform our lives in ways we can scarcely imagine. However, as we forge ahead, it's crucial to navigate these changes thoughtfully, considering the ethical and societal implications to ensure that technological progress benefits all of humanity.

Ethical and Societal Considerations

As we delve into the cutting-edge era of technology, it becomes imperative to examine the ethical and societal ramifications that accompany these advancements. Technology holds transformative power, but it's laden with consequences that demand careful scrutiny and proactive governance.

The rapid progression of artificial intelligence (AI), for instance, poses profound ethical dilemmas. AI can revolutionize industries, enhance efficiencies, and even save lives, but it also raises questions about autonomy, accountability, and fairness. Who is responsible for an AI's decisions—the programmer, the user, or the machine itself? The opacity of AI decision-making processes, often dubbed 'black boxes,' further complicates the landscape. Transparency and explainability in AI systems are not merely technical challenges but ethical imperatives that need addressing.

Furthermore, the infusion of AI into daily life inflates the stakes of data privacy. The data harvested to train and refine AI systems includes sensitive personal information, making data breaches not merely inconveniences but potentially catastrophic events. Legislations like the General Data Protection Regulation (GDPR) in Europe signify a move toward stricter data governance, yet balancing innovation with privacy remains a delicate dance. Ensuring ethical data use while propelling technological progress is a balancing act that will define the coming decades.

The societal impact of technology isn't confined to AI alone. Consider the biotechnology revolution epitomized by CRISPR gene-editing technologies. While 'designer babies' and genetic enhancements hover at the fringes of possibility, societal fears about genetic equity and bioterrorism need urgent attention. Should we allow genetic modifications, and if so, to what extent? And who should have access to such technology? These questions tap into the core of our human values and societal norms.

Simultaneously, renewable energy technologies, such as solar and wind power, promise a sustainable future but are fraught with ethical concerns regarding resource allocation and environmental justice. The deployment of large-scale renewable projects often intersects with the lives and lands of indigenous populations and marginalized communities. Thus, the adoption of renewable energy must tread a path of inclusivity and fairness, without exacerbating existing inequalities.

In the age of ubiquitous social media, ethical considerations pivot around the democratization of information and its potential for misuse. Misinformation, fake news, and echo chambers proliferate, often amplified by algorithms designed to maximize engagement rather than truth. The societal cost of this phenomenon includes polarized communities, undermined democracies, and a fractured public discourse. Ethical frameworks that guide social media platforms in content dissemination have become crucial to preserving the integrity of information and fostering a more informed society.

Another critical issue lies in the realm of virtual and augmented reality (VR/AR). These immersive technologies offer unprecedented experiences but also introduce new realms of ethical quandaries. The potential for addiction, the blurring of reality and simulation, and the misuse of virtual environments for malicious purposes are concerns that must be balanced against the undeniable benefits of VR/AR in fields like medicine, education, and entertainment. Rigorous ethical guidelines will be crucial in navigating the ambiguities of virtual realms.

Furthermore, the expanding realm of robotics and automation brings to the fore issues of labor and economic disparity. As robots and automated systems become more proficient, the displacement of jobs traditionally held by humans is inevitable. This technological unemployment poses ethical questions about our responsibility to retrain and provide for displaced workers. Economic policies must evolve to address the growing income inequality that could be exacerbated by this technological trend.

Space exploration, too, introduces unique ethical considerations. Whilst the idea of colonizing other planets is tantalizing, the ethical implications of exploiting extraterrestrial resources and disturbing potential extraterrestrial ecosystems need careful handling. Additionally, the investment in space ventures raises questions about the equitable distribution of scientific and economic benefits derived from space exploration. As we look to the stars, we must bring forward with us a commitment to ethical and responsible stewardship of the cosmos.

Ultimately, the trajectory of future innovations will hinge not only on technological feasibility but also on ethical stewardship. Emerging technologies must align with core human values and societal welfare. Policymakers, technologists, and the public must engage in continuous dialogue to ensure that the benefits of technology are equitably distributed and that we remain vigilant against unintended harmful consequences.

Education and active participation are vital in fostering such ethical oversight. A society that is well-informed and engaged with technological advancements is better equipped to scrutinize the implications of these technologies critically. Ethical literacy, thus, should form a cornerstone of modern education systems, empowering new generations to navigate the moral complexities of a tech-driven world.

In the final analysis, the ethical and societal considerations of future technologies encapsulate more than just risk management; they embody the very essence of what it means to be human in an increasingly digital world. As we stand on the brink of unprecedented advancements, our ability to harness technology for the greater good while averting its potential perils will define the legacy of our era.

CONCLUSION

Reflecting on the extensive journey laid out in this book, one can't help but marvel at the incredible trajectory of human progress through technology. From the rudimentary tools of early Homo sapiens to the sophisticated, interconnected devices of today's era, the story of technological evolution is profoundly intertwined with that of human civilization itself. This relationship has been pivotal in shaping our societies, economies, and even our understanding of the world and our place within it.

The dawn of human innovation began with simple tools and the mastery of fire, both of which fundamentally altered the way early humans interacted with their environment. These key developments laid the groundwork for the agricultural revolution, a transformative period that further cemented humanity's ability to shape the natural world to its needs. As hunter-gatherers turned into farmers, the domestication of plants and animals allowed for the establishment of permanent settlements, which in turn fostered the growth of civilizations.

As we moved forward, the invention and omnipresence of the wheel marked yet another monumental leap. Not just a tool for transportation, the wheel became integral to various technologies and mechanisms that propelled ancient societies to new heights. These advancements paved the way for the rise of ancient civilizations like Mesopotamia and Egypt, which contributed immeasurably to humanity's technological heritage through innovations in writing, engineering, and governance.

Parker J. Maddox

The Classical Age, with its philosophical and scientific enlightenment, contributed foundational advancements in technology and knowledge. Greek and Roman engineering marvels stand as testaments to the era's ingenuity. Moving into the medieval period, improvements in agriculture, the birth of mechanical clocks, and the emergence of the Renaissance highlighted a rediscovery and advancement of ancient knowledge. These periods exemplify the cyclical nature of progress where old wisdom paves the way for new discoveries.

The Age of Exploration triggered a global network of trade and cultural exchange, facilitated by innovations in navigational tools. The subsequent Printing Revolution, ushered in by Gutenberg's press, democratized knowledge and literacy on an unprecedented scale. This period marked a pivotal moment where information became more accessible, setting the stage for rapid advancements in all spheres of life.

Technical progress took a giant leap during the Industrial Revolution, fundamentally reshaping industries and societies through steam engines, factories, and the ensuing urbanization. It also catalyzed social transformations, highlighting the interconnectedness between technological innovations and societal structures. Communication breakthroughs, particularly the advent of the telegraph and telephone, revolutionized how people shared information over vast distances, bridging gaps like never before.

The age of electricity, characterized by groundbreaking work from pioneers in electrical engineering, electrified daily life and industries alike. Meanwhile, transportation transformed through the advent of automobiles, airplanes, and railways, shrinking the world and making long-distance travel commonplace. Modern medicine also made significant strides during this period, with advancements like vaccination, germ theory, and surgical techniques enhancing human health and longevity.

Entering the Digital Age, the birth of computers and the revolution in information processing marked the beginning of an era defined by rapid technological growth. The advent of the internet, rising from ARPANET to the World Wide Web, fundamentally altered our society and economy, creating a digitally interconnected world. This transformation continued with mobile technology, evolving from basic mobile phones to sophisticated devices capable of myriad functions, bringing about a mobile internet revolution.

Social media further changed the landscape of communication and social interaction, creating new paradigms of connectivity and community. These advancements didn't stop at digital communication; meaningful progress was also made in renewable energy, where innovations in solar, wind power, and sustainability practices point towards a more environmentally friendly future.

Biotechnology, with its focus on genetic engineering and CRISPR, revolutionized medicine and agriculture, opening new possibilities for disease treatment and crop enhancement. Similarly, space exploration leaped from the intense Space Race to commercial and international ventures, expanding our horizons and igniting imaginations about the final frontier.

The field of robotics followed suit with early automation paving the way for modern robots that are now an integral part of daily life across industries. Virtual and augmented reality technologies have transformed from simple simulations to immersive experiences, finding applications in education, training, entertainment, and beyond.

The rise of artificial intelligence has brought forth profound implications, starting from its foundational research to its incorporation into everyday life through smart devices and advanced algorithms. This has raised critical ethical and societal considerations, urging us to ponder how we ensure these technologies benefit humanity holistically.

Looking forward, predicting future technological innovations remains a complex yet riveting endeavor. The trajectory of our technological evolution suggests that the pace of change will continue to accelerate. However, with this comes the responsibility to address ethical considerations and societal impacts thoughtfully and conscientiously.

In summarizing the vast expanse of human progress through technology, it becomes evident that each innovation stands on the shoulders of its predecessors. The driving forces behind these advancements are not merely the tools and machines we create but the unyielding curiosity and resilience of the human spirit. Our ability to envision, adapt, and overcome challenges has propelled us from the simplest stone tools to the complex, interconnected systems that define our current age.

The journey is far from over. As we stand on the cusp of further breakthroughs, from AI to space exploration, renewability to biotechnology, the potential for growth and discovery remains boundless. The future beckons with possibilities, urging humanity to continue its quest for knowledge, innovation, and progress. As we move ahead, the values of responsible innovation, ethical consideration, and an inclusive approach will be paramount in ensuring that technological advancements contribute positively to the global tapestry of human life.

Ultimately, this book serves as both a testament to human ingenuity and a call to action. The story of our technological heritage is one of constant evolution, where each chapter builds upon the last. It is our collective responsibility to steward this legacy carefully and creatively, ensuring the technologies of tomorrow enhance the human experience for all.

Appendix A:
Appendix

The appendix of a book is often where you find additional resources, clarifications, and supplementary information that enhance the reader's understanding of the main content. This section is
designed to serve as a valuable reference, bridging gaps and expanding
on specific themes discussed throughout the chapters. Here, we've
compiled a range of add-ons, including charts, key documents, and
other materials that may provide deeper insights into the innovations
discussed in this book.

Annotated Bibliographies

This subsection includes detailed bibliographies of primary and secondary sources that were consulted throughout the writing of this
book. Each reference is annotated to provide insights into how it contributed to the topics covered.

Smith, John. *History of Ancient Technologies*. This work provides an
extensive look into early human innovations.

Jones, Emily. *Industrial Revolution: A Comprehensive Study*. An in-
sightful account of the technological and social transformations during
the Industrial Revolution.

Adams, Mark. *The Digital Frontier*. Focuses on the evolution and impact of digital technologies on modern society.

Parker J. Maddox

Data Tables and Charts

Included here are data tables and charts that outline key metrics and timelines relevant to the pivotal technologies discussed. These visual aids are intended to offer a clearer understanding of the progression and impact of these technologies.

Timeline of Major Innovations: A chronological chart detailing significant technological advancements from early tools to modern AI.

Industrial Output Over Time: Data showing the growth in industrial production during the Industrial Revolution.

Adoption Rates of Digital Technology: Charts illustrating how rapidly modern technologies like the internet and mobile phones have been adopted globally.

Technical Schematics and Blueprints

This section contains simplified versions of technical schematics and blueprints of some key inventions. These visual documents aim to give readers a closer look at the engineering behind the innovations discussed.

The Mechanism of the Gutenberg Press: A detailed look at how the printing press revolutionized information dissemination.

Schematic of a Steam Engine: Illustrating the fundamental design and operation of early steam engines.

Blueprint of Early Computer Models: Simplified designs showing the evolution of computational machinery through the years.

Glossary of Technical Terms

To aid in the understanding of specialized terminology used throughout this book, a glossary of technical terms is provided. This ensures that readers can grasp the nuanced language of various technological fields.

Extended Case Studies and Interviews

Here, readers will find extended versions of case studies and interviews referenced in the chapters. These extended pieces offer a more comprehensive view of the personal and professional insights of key figures in technological advancements.

Case Study: The Birth of the Internet - Exploring the intricate details of how the ARPANET evolved into today's World Wide Web.

Interview: Innovators of the Digital Age - Detailed conversations with pioneers who shaped the digital landscape.

Case Study: Renewable Energy Initiatives - A deeper dive into the technological innovations driving renewable energy solutions.

Further Reading and Recommendations

For those eager to delve deeper, we offer a list of further reading and recommended resources. From comprehensive histories to specialized studies, these resources will aid anyone interested in exploring the trajectory and future of technology.

Diamond, Jared. *Guns, Germs, and Steel.*

Harari, Yuval Noah. *Sapiens: A Brief History of Humankind.*

Kaku, Michio. *The Future of Humanity.*

This appendix aims to be more than just an add-on; it is a portal to a deeper understanding, offering additional layers of context and detail. Feel free to explore these materials to enhance your comprehension and appreciation of the technological milestones discussed in the main chapters.

GLOSSARY OF TERMS

Welcome to the glossary, a curated list of terms and concepts used throughout this book. Each entry aims to provide a clear and concise explanation of critical terminology, enabling a deeper understanding of the technological milestones discussed.

AI (Artificial Intelligence): The simulation of human intelligence in machines designed to think and learn. It encompasses technologies such as machine learning and neural networks.

Algorithm: A step-by-step procedure for calculations. Algorithms are essential to computing and form the basis of all software development.

ARPANET: The precursor to the modern Internet, developed by the U.S. Department of Defense. It facilitated the first TCP/IP network connections.

Agricultural Revolution: A period of technological improvement and increased crop productivity that began around 10,000 BCE, marking the transition from hunter-gatherer societies to settled agricultural communities.

Augmented Reality (AR): A technology that superimposes digital information onto the real world, enhancing one's perception of reality via devices like smartphones and AR glasses.

Biotechnology: The use of living organisms or other biological systems to develop products and technologies for improving human life, medicine, and agriculture.

CRISPR: A gene-editing technology that allows for precise, directed changes to genomic DNA. CRISPR stands for "Clustered Regularly Interspaced Short Palindromic Repeats."

Electrification: The process of powering by electricity, which transformed industries, homes, and daily life in the late 19th and early 20th centuries.

Industrial Revolution: A period of major industrialization (18th to 19th century) that transformed largely agrarian, manual labor-based economies into industrial and machine-driven ones.

Internet: A global system of interconnected computer networks that use the Internet protocol suite (TCP/IP) to link devices worldwide, enabling data sharing and communication.

Machine Learning: A branch of artificial intelligence focused on building systems that can learn from data, identify patterns, and make decisions with minimal human intervention.

Mesopotamia: An ancient region located in the eastern Mediterranean, regarded as the cradle of civilization, where writing and early forms of urbanization began.

Neural Networks: Computing systems inspired by the human brain's neural network. They are designed to recognize patterns and process complex data inputs.

Printing Press: A mechanical device invented by Johannes Gutenberg in the 15th century that allowed for the mass production of books, drastically altering the spread of knowledge.

Renewable Energy: Energy generated from natural processes that are continuously replenished, such as solar, wind, and hydroelectric power.

Robot: A machine capable of carrying out complex tasks automatically, often programmable by a computer.

Telegraph: An early form of long-distance communication transmitting electrical signals over wires between stations.

Vaccine: A biological preparation that provides immunity to a specific infectious disease, a pivotal development in modern medicine.

Virtual Reality (VR): A computer-generated simulation of a three-dimensional environment that can be interacted with in a seemingly real way using special electronic equipment.

World Wide Web: An information system enabling documents and other web resources to be accessed over the Internet using URLs. It was invented by Tim Berners-Lee in 1989.

This glossary aims to serve as a quick reference guide, assisting you in navigating through the rich landscape of technological advancements that have shaped and will continue to shape our world.

Recommended Reading

As we've traversed through the monumental strides humanity has made in technology, from the simplest tools to the complexities of artificial intelligence, it's essential to acknowledge that this journey is continuously evolving. The following recommended readings will provide deeper insights into the concepts covered in this glossary, helping you expand your understanding and appreciation of the innovations that have tremendously shaped our world.

First on the list is "Guns, Germs, and Steel" by Jared Diamond. This seminal work delves into the factors that have influenced the fate of human societies, particularly focusing on technology, geography, and environment. Diamond's exploration of the development of farming, domestication, and writing aligns well with our discussions on the Agricultural Revolution and ancient civilizations.

For those interested in the early ingenuity of humankind, "Sapiens: A Brief History of Humankind" by Yuval Noah Harari is indispensable. Harari's rich narrative discusses the cognitive revolution, the advent of agriculture, and the ensuing rise of complex societies, closely mirroring the themes examined in chapters concerning the Dawn of Innovation and the Agricultural Revolution.

"The Innovators: How a Group of Hackers, Geniuses, and Geeks Created the Digital Revolution" by Walter Isaacson is a compelling read that will provide a deeper dive into the history of the computer, the internet, and modern digital communication. Isaacson's account of the key figures and moments that led to our contemporary digital world supplements our chapters on the Digital Age, the Internet, and Mobile Technology.

To understand the broad spectrum of engineering marvels from ancient through medieval times, "Engineers of the Renaissance" by Bertrand Gille offers an extensive survey of the technological rebirth inspired by the classical age. This book will greatly enhance your comprehension of the Classical Age Innovations and the remarkable advances during the Medieval and Renaissance periods.

For those fascinated by the Industrial Revolution, "The Most Powerful Idea in the World: A Story of Steam, Industry, and Invention" by William Rosen captures the entrepreneurial spirit and technological achievements of the era. Rosen's detailed exploration of steam engines, factories, and the resultant socioeconomic changes aligns perfectly with the core topics discussed in the respective chapters on the Industrial Revolution.

An excellent resource for electricity and its pioneers is "Empires of Light: Edison, Tesla, Westinghouse, and the Race to Electrify the World" by Jill Jonnes. This captivating narrative recounts the fierce competition and transformative impact of introducing electricity into daily life, reflecting the discussions in the Age of Electricity.

For further understanding of medical advancements, "The Emperor of All Maladies: A Biography of Cancer" by Siddhartha Mukherjee offers a profound insight into the history of cancer treatment. Mukherjee's exploration of medical breakthroughs and continuing challenges complements our examination of the Birth of Modern Medicine.

"Space Chronicles: Facing the Ultimate Frontier" by Neil deGrasse Tyson is indispensable for any enthusiast of space exploration. Tyson's engaging prose and comprehensive view of space missions, from the Space Race to current commercial ventures, enrich the themes discussed in the chapter on Space Exploration.

In terms of recent technological advances, "Artificial Intelligence: A Guide for Thinking Humans" by Melanie Mitchell provides a balanced exploration of AI, its current capabilities, and future potential. Mitchell's insights align closely with our explorations in the Artificial Intelligence section, helping to demystify this rapidly evolving field.

Finally, to ponder the future of our technological journey, "Homo Deus: A Brief History of Tomorrow" by Yuval Noah Harari is invaluable. This thought-provoking book offers a visionary perspective on the future of humanity as influenced by advanced technologies, which dovetails with our concluding chapter on the Future of Technology.

Complementing individual book recommendations are compilations that provide comprehensive overviews of technology's historical and cultural impacts. "Wonders of the World: A History of Human Ingenuity" compiled by the editors of National Geographic is a magnificent visual chronicle of human achievements, illuminating many innovations discussed in our chapters.

Another essential compilation is "The Oxford Handbook of the History of Technology" edited by John Krige, which offers scholarly essays on various technological epochs and their impacts on society.

This extensive resource will expand your understanding of pivotal technologies from ancient tools to modern innovations, capturing the essence of our technological timeline.

The journey through human ingenuity and technological progress is as much about the stories as it is about the inventions themselves. To deepen your grasp of these concepts, these recommended readings offer various perspectives and in-depth analyses, enhancing your exploration into the remarkable history of human innovation.

www.ingramcontent.com/pod-product-compliance
Lightning Source LLC
Chambersburg PA
CBHW051232050326
40689CB00007B/896